The North & West Midlands Bus Handbook

Counties in this issue:

Shropshire

Staffordshire

West Midlands

April 1996

British Bus Publishing

The North & West Midlands Bus Handbook

The North & West Midlands Bus Handbook is part of the Bus Handbook series that details the fleets that are used by stage carriage and express coach operators. Where space allows other significant operators in the area are also included. These handbooks are published by *British Bus Publishing* and cover Scotland, Wales and England north of London. The current list is shown at the end of the book. Together with similar books for southern England, published by Capital Transport which we also supply, they provide comprehensive coverage of all the principal operators' fleets in the British Isles. Handbooks for the FirstBus Group and Stagecoach are also published annually.

The fleets included in this edition are those for operators based, and who provide stage and express services, in the counties of Shropshire, Staffordshire and the West Midlands. Also included are a number of those operators who provide significant coaching activities.

Quality photographs for inclusion in the series are welcome, for which a fee is payable. The publishers unfortunately cannot accept responsibility for any loss and request you show your name on each picture or slide. Details of changes to fleet information are also welcome.

To keep the fleet information up to date we recommend the Ian Allan publication *Buses* published monthly, or for more detailed information, the PSV Circle monthly news sheets.

The writer and publisher would be glad to hear from readers should any information be available which corrects or enhances that given in this publication.

Series Editor: Bill Potter
Principal Editors for *The North & West Midlands Bus Handbook*:
Bill Potter & David Donati

Acknowledgements:
We are grateful to Andy Chown, Tony Hunter, Mark Jameson, Colin Lloyd, Steve Sanderson the PSV Circle and the operating companies for their assistance in the compilation of this book.

The front cover photograph is by Barrie Kelsall

Contents correct to April 1996

ISBN 1 897990 04 9
Published by *British Bus Publishing Ltd*
The Vyne, 16 St Margarets Drive, Wellington,
Telford, Shropshire, TF1 3PH
© British Bus Publishing, April 1996

CONTENTS

Bakers	4	Leon's of Stafford	46
Banga Travel	6	Lionspeed / Pete's	49
Bassetts	7	Little Red Bus	50
The Birmingham Coach		Longmynd	51
Company	9	Ludlows	52
Blue Buses	12	M&J Travel	53
Boultons of Shropshire	13	Merry Hill Mini	54
Bowens	15	Metropolitan	55
Boydon	16	Midland	56
Britannia	17	Minsterley Motors	65
Butters	19	Moorland Buses	66
Cave	20	N C B	67
Chase Bus Services	21	NCP	68
Choice Travel	24	North Birmingham	69
Claribels	27	Owen's	70
Clowes	27	Patterson	71
Copelands	28	PMT	72
Elcock Reisen	29	Procters	80
Falcon Travel	31	Rest & Ride	81
Flights	33	Sandwell Travel	82
Glenstuart Travel	34	Serveverse	83
Green Bus Service	35	Shropshire Education	84
Handybus	37	Stevensons	85
Happy Days	38	Travel De Courcey	92
Harry Shaw	40	Warringtons	94
Hi Ride	40	W M Buses	95
Horrocks	43	Williamsons	113
Jones	43	Worthen Travel	114
King Offa	44	Zak's	115
Knotty	44	Index of vehicles	116

BAKERS

Guideissue Ltd, Spring Grove, Congleton Road, Biddulph, Staffordshire ST8 7RQ

Depots : Wallis Street, Biddulph and Highway Garage, Rudyard

1	5658RU	Volvo B10M-62	Plaxton Premiére 350	C49FT	1994	
2	9530RU	Volvo B10M-62	Plaxton Premiére 350	C49FT	1994	
3	7092RU	Volvo B10M-61	Plaxton Paramount 3200 III	C53F	1988	
4	3601RU	Volvo B10M-62	Plaxton Premiére 320	C57F	1996	
5	D780FVT	Volvo B10M-61	Plaxton Paramount 3200 III	C49FT	1987	Ex Excelsior, 1990
6	3275RU	Volvo B10M-60	Plaxton Paramount 3200 III	C53F	1990	Ex Excelsior, 1991
7	1513RU	Volvo B10M-62	Plaxton Premiére 320	C57F	1996	
8	8150RU	Volvo B10M-60	Plaxton Premiére 350	C49FT	1992	Ex Redwing, Camberwell, 1995
9	1497RU	Volvo B10M-60	Plaxton Premiére 350	C49FT	1992	Ex Redwing, Camberwell, 1995
11	9995RU	Volvo B10M-60	Plaxton Paramount 3500 III	C49FT	1991	Ex Ambassador, 1996
12	8399RU	Volvo B10M-60	Plaxton Paramount 3500 III	C49FT	1991	Ex Ambassador, 1996
15	1879RU	Volvo B10M-62	Plaxton Premiére 320	C49FT	1994	Ex Excelsior, 1996
17	3353RU	Volvo B10M-62	Plaxton Premiére 320	C49FT	1994	Ex Excelsior, 1996
18	3471RU	Mercedes-Benz L307D	?	M12	1994	
19	6577RU	Mercedes-Benz 410D	Autobus Classique	M16	1992	
20	3102RU	Mercedes-Benz 711D	Dormobile Routemaker	C29F	1993	Ex Holloway, Scunthorpe, 1994
21	9595RU	Mercedes-Benz L307D	Imperial	M12	1985	Ex Clapton Coaches, 1994
22	K6BUS	Mercedes-Benz 811D	Dormobile Routemaker	B33F	1992	Ex Patterson, Birmingham, 1995
23	K7BUS	Mercedes-Benz 811D	Dormobile Routemaker	B33F	1992	Ex Patterson, Birmingham, 1995
27	9423RU	Mercedes-Benz 811D	Optare StarRider	C29F	1989	
28	5777RU	Mercedes-Benz 609D	Whittaker Europa	C21F	1990	
29	5621RU	Mercedes-Benz 609D	Whittaker Europa	C21F	1990	
30	3093RU	Volvo B10M-60	Plaxton Paramount 3500 III	C53F	1989	Ex Park's, 1990
31	7025RU	Volvo B10M-60	Plaxton Paramount 3500 III	C53F	1989	Ex Park's, 1990
32	6280RU	Volvo B10M-60	Plaxton Paramount 3500 III	C49FT	1989	Ex Park's, 1991
33	3563RU	Volvo B10M-60	Plaxton Paramount 3500 III	C49FT	1989	Ex Park's, 1991
34	3566RU	Volvo B10M-60	Plaxton Paramount 3500 III	C38FT	1989	Ex Scarborough FC, 1991
35	8830RU	Mercedes-Benz 811D	Optare StarRider	C29F	1992	

Previous Registrations:

1497RU	J435HDS	3566RU	F684EAG	8150RU	J434HDS
1513RU	From new	3601RU	From new	8399RU	H172EJU
1879RU	XEL55C, L327ERU	5621RU	G429NET	8830RU	From new
3093RU	F36HGG	5658RU	From new	9423RU	From new
3102RU	K347EJV	5777RU	G428NET	9530RU	From new
3275RU	G512EFX	6280RU	F977HGE	9595RU	B374WHY
3353RU	XEL6C, L326ERU	6577RU	From new	9995RU	H171EJU
3471RU	From new	7025RU	F34HGG, 7025RU, F29WRE	D780FVT	D255HFX, 4614RU
3563RU	F987HGE	7092RU	From new	F940WFA	F487LHO, 3353RU

Illustrating the Bakers fleet are minibuses K7BUS, one of two with Dormobile Routemaker, and 5621RU, with Whittaker Europa bodywork seen here in China Link livery for the service linking various potteries. Representing the main coach fleet is 9530RU a Volvo B10M with Plaxton Premiére 350 bodywork.
Cliff Beeton/ Roy Marshall

C1 CHINA LINK
HOURLY SERVICE AROUND POTTERY FACTORY SHOPS
CHINA LINK
BAKERS
5621 RU

BAKERS
BAKERS Eurocruiser
VOLVO
PLAXTON
9530 RU

BANGA TRAVEL

P R Banga, 4 Vicarage Road, Wolverhampton WV2 1DT

Depots: Cannock Road Garage, Wolverhampton and Hickman Avenue, Wolverhampton

B1	D69OKG	Freight Rover Sherpa	Carlyle	B18F	1987	Ex Delivered-in-Style, Syston, 1993
B2	D528NDA	Freight Rover Sherpa	Carlyle	B18F	1986	Ex Williams, Runcorn, 1993
B3	D190NON	Freight Rover Sherpa	Carlyle	B18F	1987	Ex MCH, Uxbridge, 1993
B4	E410EPE	Renault-Dodge S46	Northern Counties	B22F	1987	Ex Stagecoach South, 1993
B7	D31SAO	Renault-Dodge S56	Reeve Burgess	B23F	1986	Ex Stagecoach South, 1994
B8	D926KWW	Renault-Dodge S56	Northern Counties	B22F	1987	Ex Harrogate & District, 1995
B9	D925KWW	Renault-Dodge S56	Northern Counties	B22F	1987	Ex Harrogate & District, 1995
	D63MTG	Renault-Dodge S56	East Lancashire	DP22F	1987	Ex Victoria Travel, Earlestown, 1994
	C503PSC	Renault-Dodge S56	Dormobile	B26F	1986	Ex Symons, Motherwell, 1994
	C505PSC	Renault-Dodge S56	Dormobile	B26F	1986	Ex Symons, Motherwell, 1994
	D302MHS	Renault-Dodge S56	Alexander AM	B21F	1987	Ex Kingscliff Minibuses, Methlick, 1995
	D109OWG	Renault-Dodge S56	Reeve Burgess	B25F	1986	Ex Victoria Travel, Earlestown, 1995

Livery: Cream, black and blue

Pictured passing along Lichfield Street in Wolverhampton is C505PSC, one of a pair of Dormobile-bodied Renault-Dodge S56s acquired by Banga Travel in 1994. *Tony Wilson*

BASSETTS

Bassetts Coachways Ltd, Transport House, Tittensor, Staffordshire ST12 9HD

NWW163K	Bristol LH6L	Plaxton Elite II	C45F	1972	Ex Robinson, Great Harwood, 1977
KCW74N	Leyland Leopard PSU5/4R	Duple Dominant	C57F	1975	Ex Robinson, Great Harwood, 1981
KCW75N	Leyland Leopard PSU5/4R	Duple Dominant	C57F	1975	Ex Robinson, Great Harwood, 1980
RFR177P	Leyland Leopard PSU3C/4R	Duple Dominant	C51F	1976	Ex Robinson, Great Harwood, 1981
JRE354V	Leyland Leopard PSU3E/4R	Plaxton Supreme IV Express	C51F	1979	Ex Middleton, Rugeley, 1981
WVH868V	Leyland Leopard PSU3E/4R	Duple Dominant II Express	C53F	1979	Ex Longstaff, Mirfield, 1983
LVS423V	Leyland Leopard PSU5C/4R	Duple Dominant II	C57F	1980	Ex Ebdon, Sidcup, 1983
XGS764X	Leyland Tiger TRCTL11/3R	Plaxton Supreme IV	C57F	1981	Ex Ebdon, Sidcup, 1983
XBF423X	Leyland Tiger TRCTL11/2R	Plaxton Supreme V Express	C53F	1982	
VJT607X	Ford R1114	Plaxton Supreme V	C49F	1982	Ex Staffordshire Police, 1994
EBF806Y	Leyland Tiger TRCTL11/3R	Plaxton Paramount 3200	C57F	1983	
E93MRF	Mercedes-Benz 609D	Reeve Burgess Beaver	C25F	1988	
E542MRE	Ford Transit VE6	Dormobile	M16	1988	
F513RTL	Dennis Javelin 12SDA1908	Plaxton Paramount 3200 III	C48FT	1989	Ex Grimsby-Cleethorpes, 1994
F877RFP	Dennis Javelin 12SDA1907	Duple 320	C57F	1989	Ex Clarkes Coaches, Pailton, 1991
F878RFP	Dennis Javelin 12SDA1907	Duple 320	C57F	1989	Ex Clarkes Coaches, Pailton, 1992
F879RFP	Dennis Javelin 12SDA1907	Duple 320	C57F	1989	Ex Clarkes Coaches, Pailton, 1991
G472EFA	Dennis Javelin 12SDA1907	Duple 320	C57F	1990	
G278BEL	Dennis Javelin 11SDA1906	Duple 320	C53F	1989	Ex Luckett, Fareham, 1992
H922FEW	Iveco Daily 49.10	Videofit	M15L	1990	Ex Cambridgeshire CC, 1995
J4MMT	Dennis Javelin 11SDL1921	Plaxton Paramount 3200 III	C53F	1992	Ex McLaughlin, Penwortham, 1993
J842NNR	Dennis Javelin 12SDA1929	Plaxton Paramount 3200 III	C53F	1992	Ex Gardner, Upton, 1994
K440DVT	Mercedes-Benz 814D	Autobus Classique	C33F	1993	
L962NFA	Dennis Javelin 12SDA2131	Plaxton Premiére 320	C57F	1994	
M881WFA	Dennis Javelin 12SDA2131	Plaxton Premiére 320	C53F	1995	
M882YEH	Dennis Javelin 12SDA2155	Plaxton Premiére 320	C53F	1995	
N755GBF	Mercedes-Benz 711D	Reeve Burgess Beaver	C25F	1996	

Previous Registrations:
F513RTL   F638UBE,PSU443

Livery:
Pale blue and grey

The Dennis Javelin has been taken into the Bassetts fleet regularly as part of its modernisation programme since the first example was supplied new in 1990. That Duple-bodied example has since been joined by three further new coaches and six previously used elsewhere. The latest arrival is M882YEH and carries a Plaxton Premiére 320 body. *Cliff Beeton*

9 HALESOWEN
The Birmingham Coach Company
WE GIVE CHANGE
The Birmingham Coach Company
83
PTF 730L

DUDLEY 120
The Birmingham Coach Company
WE GIVE CHANGE
The Birmingham Coach Company
J56 GCX
VBU 945X

THE BIRMINGHAM COACH COMPANY

Birmingham Omnibus Co Ltd; Birmingham Coach Co Ltd; Cross Quays Business Park, Hallbridge Way, Tipton Road, Tividale, West Midlands B69 3HY

2	KSO76P	Leyland National 10351/2R	B40F	1976	Ex Trefaldwyn, Montgomery, 1988
3	GUG118N	Leyland National 11351/1R (Volvo)	B52F	1974	Ex PMT, 1989
4	HJA131N	Leyland National 10351/1R	B41F	1975	Ex Ludlow, Halesowen, 1990
6	NOA199P	Leyland National 11351A/1R	B50F	1976	Ex Enterprise, Coventry, 1989
10	MTJ775S	Leyland National 11351A/1R	B49F	1977	Ex Merseybus, 1991
12	WYJ167S	Leyland National 11351A/2R	B44D	1978	Ex Brighton & Hove, 1992
14	GRM351L	Leyland National 1151/1R/0401	B52F	1972	Ex Cumberland, 1990
15	GRM353L	Leyland National 1151/1R/0401	B52F	1973	Ex Cumberland, 1990
16	GEU369N	Leyland National 10351/1R	B44F	1974	Ex Shearings, 1989
17	GEU371N	Leyland National 10351/1R	B44F	1975	Ex Shearings, 1989
19	HHU633N	Leyland National 10351/1R (Volvo)	B44F	1974	Ex Shearings, 1989
20	HHU634N	Leyland National 10351/1R	B44F	1975	Ex Shearings, 1989
22	GHU641N	Leyland National 10351/1R	B44F	1975	Ex Badgerline, 1986
23	BSF766S	Leyland National 11351A/1R	B52F	1977	Ex Western Scottish, 1991
25	RYG768R	Leyland National 11351A/1R	B52F	1976	Ex West Yorkshire, 1988
26	KOM789P	Leyland National 11351/2R	B46D	1976	Ex Shearings, 1989
28	KOM793P	Leyland National 11351/2R	B46D	1976	Ex Shearings, 1989
30	NEL860M	Leyland National 1151/1R/2402	B49F	1974	Ex Shearings, 1989
31	MOD826P	Leyland National 11351A/1R	B50F	1976	Ex Shearings, 1989
33	YCW843N	Leyland National 10351/1R	B44F	1974	Ex Shearings, 1989
34	YCW845N	Leyland National 10351/1R	B44F	1974	Ex Shearings, 1989
35	MOD850P	Leyland National 11351A/1R	B50F	1976	Ex Shearings, 1989
37	SKF18T	Leyland National 11351A/1R	B49F	1979	Ex Merseybus, 1991
40	EGB91T	Leyland National 11351A/1R	B52F	1979	Ex Western Scottish, 1991
41	EGB90T	Leyland National 11351A/1R	B52F	1979	Ex Western Scottish, 1991
44	SKF13T	Leyland National 11351A/1R	B49F	1979	Ex Merseybus, 1991
45	RKA873T	Leyland National 11351A/1R	B49F	1978	Ex Merseybus, 1991
46	RKA877T	Leyland National 11351A/1R	B49F	1978	Ex Merseybus, 1991
47	RKA878T	Leyland National 11351A/1R	B49F	1978	Ex Merseybus, 1991
48	RKA884T	Leyland National 11351A/1R	B49F	1978	Ex Merseybus, 1991
49	SKF20T	Leyland National 11351A/1R	B52F	1979	Ex Merseybus, 1992
50	SKF30T	Leyland National 11351A/1R	B52F	1979	Ex Merseybus, 1992
56	YPF774T	Leyland National 10351A/1R	B41F	1978	Ex Reeves, Withnell, 1993
62	UFG55S	Leyland National 11351A/2R	B44D	1977	Ex Brighton & Hove, 1992
63	UFG57S	Leyland National 11351A/2R	B44D	1977	Ex Brighton & Hove, 1992
64	AYJ106T	Leyland National 11351A/2R	B52F	1979	Ex Brighton & Hove, 1992
65	YCD80T	Leyland National 11351A/2R	B44D	1978	Ex Brighton & Hove, 1992
66	UFG59S	Leyland National 11351A/2R	B44D	1977	Ex Brighton & Hove, 1992
67	WYJ166S	Leyland National 11351A/2R	B44D	1978	Ex Brighton & Hove, 1992
68	WPG224M	Leyland National 10351/1R/SC	DP39F	1974	Ex Cowes Express, IoW, 1992
69	LPR940P	Leyland National 11351/1R	B49F	1976	Ex WMRCC, Swanley, 1992
71	HSC106T	Leyland National 11351A/1R	B49F	1978	Ex Fife Scottish, 1992
74	YCD79T	Leyland National 11351A/2R	B44D	1978	Ex Brighton & Hove, 1992
75	AKU160T	Leyland National 10351B/1R	B44F	1979	Ex Aintree Coachline, 1993
76	YYE291T	Leyland National 10351A/2R	B44F	1979	Ex Cyril Evans, Sengherydd, 1993
77	THX219S	Leyland National 10351A/2R	B44F	1978	Ex Cyril Evans, Sengherydd, 1993
78	OJD862R	Leyland National 10351A/2R	B44F	1977	Ex Cyril Evans, Sengherydd, 1993
79	HMA565T	Leyland National 10351B/1R	B44F	1978	Ex Kinch, Barrow-on-Soar, 1993
80	XAK451T	Leyland National 11351A/1R	B52F	1978	Ex Clyde Coast Services, 1993
81	GMB661T	Leyland National 10351B/1R	B44F	1978	Ex Kinch, Barrow-on-Soar, 1993
82	GMB650T	Leyland National 10351B/1R	B44F	1978	Ex Kinch, Barrow-on-Soar, 1993
83	PTF730L	Leyland National 1151/2R/0401	B52F	1972	Ex Ribble, 1993
84	NTC604M	Leyland National 1151/1R/0401	B49F	1973	Ex Ribble, 1993
86	UHG745R	Leyland National 11351A/1R	B49F	1976	Ex Ribble, 1993
87	SCK708P	Leyland National 11351A/1R	B49F	1976	Ex Ribble, 1993

The Birmingham Coach Company's main area of operation lies to the south of Birmingham. Here are seen contrasting types. No.83, PTF730L, is one of the early Leyland Nationals and one of four acquired from Ribble in 1993. Since 1994, additional buses do not display fleet numbers. Falling into this category is J56GCX, one of three DAF SB220s with Ikarus bodywork. Principally imported into the UK on DAF chassis, the Citibus model is known as the 435 at home in Budapest. *Tony Wilson*

The North & West Midlands Bus Handbook

NTC614M	Leyland National 1151/1R		B49F	1973	Ex Quickstep, 1994
JOX506P	Leyland National 11351A/1R		B49F	1976	Ex Stagecoach Midland Red, 1995
NOE567R	Leyland National 11351A/1R		B49F	1976	Ex Stagecoach Midland Red, 1995
NOE568R	Leyland National 11351A/1R		B49F	1976	Ex Stagecoach Midland Red, 1995
NOE570R	Leyland National 11351A/1R		B49F	1977	Ex Stagecoach Midland Red, 1995
NOE579R	Leyland National 11351A/1R		B49F	1976	Ex Stagecoach Midland Red, 1995
222UPD	Leyland Leopard PSU3D/4R	Duple 320(1988)	C53F	1977	Ex Alexander, Sheffield, 1986
BAL607T	Leyland National 11351A/1R		B49F	1978	Ex MTL, 1995
BAL609T	Leyland National 11351A/1R		B49F	1978	Ex MTL, 1995
DAR132T	Leyland National 11351A/1R		B49F	1979	Ex City Buslines, Birmingham, 1995
SKF17T	Leyland National 11351A/1R		B49F	1979	Ex MTL, 1995
TTC539T	Leyland National 11351A/1R		B49F	1979	Ex MTL, 1995
AKY612T	Leyland National 11351A/1R (Volvo)		B49F	1979	Ex MTL, 1995
AAL272A	Leyland Leopard PSU5D/4R	Plaxton P'mount 3200(1987)	C53F	1980	Ex National Welsh, 1992
AAL345A	Leyland Leopard PSU5D/4R	Plaxton P'mount 3200(1987)	C57F	1980	Ex National Welsh, 1992
AAL453A	Leyland Leopard PSU5D/4R	Plaxton P'mount 3200(1987)	C53F	1980	Ex National Welsh, 1992
JDZ4898	Auwaerter Neoplan N122/3	Auwaerter Skyliner	CH57/20CT	1985	Ex Trathens, Plymouth, 1993
SIB8242	Volvo B10M-61	Van Hool Alizée	C49FT	1987	Ex Cambridge Coach Services, 1991
F161DET	Scania K112CRB	Van Hool Alizée	C53F	1988	Ex Team Travel, Horsforth, 1994
H10WLE	Scania K113TRB	Van Hool Astrobel	CH53/14CT	1990	Ex Busways, 1994
H682FCU	Scania K113TRB	Van Hool Astrobel	CH53/14CT	1990	Ex Busways, 1994
J413NCP	DAF SB220LC550	Ikarus Citibus	B48F	1992	Ex Richards, Cardigan, 1993
J414NCP	DAF SB220LC550	Ikarus Citibus	B48F	1992	Ex Yorkshire Travel, Wakefield, 1994
J56GCX	DAF SB220LC550	Ikarus Citibus	B48F	1992	Ex Strathclyde, 1994
K657BOH	Volvo B10M-60	Plaxton Expressliner II	C46FT	1993	On loan from Central Coaches
K658BOH	Volvo B10M-60	Plaxton Expressliner II	C46FT	1993	On loan from Central Coaches
K659BOH	Volvo B10M-60	Plaxton Expressliner II	C46FT	1993	On loan from Central Coaches
K660BOH	Volvo B10M-60	Plaxton Expressliner II	C46FT	1993	On loan from Central Coaches
M784SOF	Scania K113CRB	Van Hool Alizée	C49FT	1995	
M785SOF	Scania K113CRB	Van Hool Alizée	C49FT	1995	
M237SOJ	Scania K113CRB	Van Hool Alizée	C49FT	1995	
N683AHL	Scania K113CRB	Van Hool Alizée	C49FT	1995	
N684AHL	Scania K113CRB	Van Hool Alizée	C49FT	1995	

Previous Registrations:

222UPD	VDH243S	H682FCU	H133ACU, KSU463	K659BOH	K3CEN
AAL272A	BUH221V	JDZ4898	B668DVL	K660BOH	K2CEN
AAL345A	BUH220V	K657BOH	K5CEN	NOA199P	MOD824P, CSU993
AAL453A	BUH224V	K658BOH	K4CEN	SIB8342	D848KVE
H10WLE	H134ACU, KSU464, H681FCU				

Livery: Cream and red (buses), red, yellow and white (coaches).

Three former National Welsh Leyland Leopards with 1987 Plaxton Paramount bodywork were acquired in 1992 and have retained the index marks gained on re-bodying. This 1996 picture shows AAL272A passing the Midland Bank in Victoria. An interesting addition is the school bus sign off-side of the destination box.
Colin Lloyd

The mainstay of the bus service fleet is the Leyland National with most versions of the mark 1 represented. Seen in Walsall on service 310 to the Merry Hill shopping centre is dual-doored 12, WYJ167S proudly advising passengers that The Birmingham Coach Company gives change.
Cliff Beeton

The Birmingham Coach Company operate diagrams on several National Express services including the West Midlands to Gatwick, Coventry to Blackpool and London to Holyhead. Pictured here is K659BOH, a Volvo B10M-60 with a Plaxton Expressliner 2 body on extended loan from W M Buses' Central Coaches division liveried for Rapide duties. *Colin Lloyd*

BLUE BUSES

Scragg's Coaches & Taxis, Bucknall Garage, Pennell Street, Bucknall,
Stoke-on-Trent   ST2 9BD

	1672VT	Bedford YMQ	Plaxton Paramount 3200	C45F	1983	Ex Evans, Tregaron, 1988
	1655VT	Mercedes-Benz 507D	Reeve Burgess	M16	1987	Ex Brown, Bathgate, 1991
63	E163TWO	Freight Rover Sherpa	Carlyle Citybus 2	B20F	1988	Ex National Welsh, 1991
69	E969SVP	Freight Rover Sherpa	Carlyle Citybus 2	B20F	1987	Ex National Welsh, 1991
96	E196UKG	Freight Rover Sherpa	Carlyle Citybus 2	B20F	1988	Ex National Welsh, 1991
97	E197UKG	Freight Rover Sherpa	Carlyle Citybus 2	B20F	1988	Ex National Welsh, 1991
97	E700HLB	Mercedes-Benz 709D	Reeve Burgess Beaver	C23F	1988	Ex Globe, Barnsley, 1995
	9685VT	Bedford YNT	Duple 320	C53F	1987	Ex Ashford Luxury Coaches, 1993
	JIL5227	Mercedes-Benz 609D	Reeve Burgess Beaver	B20F	1988	Ex City Nippy, Middleton, 1995
	??	Mercedes-Benz 507D	Made-to-Measure	C20F	1988	
78	G678XVT	Mercedes-Benz 609D	Made-to-Measure	C24F	1989	
79	G279HDW	Freight Rover Sherpa	Carlyle Citybus 2	B20F	1990	Ex National Welsh, 1992
	G675BFA	Mercedes-Benz 609D	North West CS	C24F	1990	Ex Graham's, Talke, 1993
	H304HVT	Mercedes-Benz 709D	PMT Ami	DP29F	1990	

Previous Registrations:

1655VT	E222LBV	JIL5227	F377UCP
1672VT	DLL47Y	??	E416KBF, VOI3577, 6727VT
9685VT	D922GRU		

Livery: Blue, or blue and white (coaches)

Seen on lay-over in Hanley bus station is Blue Buses' E163TWO, a Freight Rover Sherpa with a Carlyle Citybus 2 body and one of five similar vehicles in the fleet previously with National Welsh.
Richard Godfrey

BOULTONS OF SHROPSHIRE

MJ and CM Boulton, Sunnyside, Cardington, Church Stretton,
Shropshire   SY6 7HR

HVJ203	Bedford OB	Duple Vista	C29F	1951	Ex Mason, Mansel Lacy, 1983
KNT780	Leyland Royal Tiger PSU1/16	Burlingham Seagull	C37C	1954	Ex preservation, 1988
JBW527D	Bedford VAM3	Duple Bella Vega	C45F	1966	Ex Gain, Westfield, 1990
ODD161M	Bedford YRT	Plaxton Elite III Express	C49F	1973	Ex Harveys, Fosse Cross, 1988
KUN497P	Bedford YRQ	Plaxton Elite III	C45F	1976	Ex Pearce, Yatton, 1989
C314NNT	Bova FLD12.250	Bova Futura	C53F	1986	Ex M&D Travel, Clive, 1991
E467VNT	Mercedes-Benz 811D	Optare StarRider	DP31F	1987	
E402YNT	Mercedes-Benz 811D	Optare StarRider	C29F	1988	Ex M&D Travel, Clive, 1996
E746JAY	Dennis Javelin 11SDL1905	Plaxton Paramount 3200 III	C53F	1988	Ex Snowdon, Easington Colliery, 1992
E749NSE	Dennis Javelin 11SDL1905	Plaxton Paramount 3200 III	C53F	1988	Ex Ipswich, 1993
E536PRU	Dennis Javelin 11SDL1905	Plaxton Paramount 3200 III	C53F	1988	Ex Tillingbourne, Cranleigh, 1991
E650JWP	Dennis Javelin 12SDA1908	Plaxton Paramount 3200 III	C49FT	1988	Ex Owen's, Oswestry, 1995
F527BUX	Mercedes-Benz 811D	Optare StarRider	DP33F	1988	
JAZ9870	Bova FHD12.290	Bova Futura	C49FT	1989	Ex M&D Travel, Clive, 1996
GIL2785	Bova FHD12.290	Bova Futura	C55F	1989	Ex The Kings Ferry, 1994
F600EAW	Ford Transit VE6	Ford	M11	1988	
G838LWR	Mercedes-Benz 811D	Optare StarRider	B31F	1990	
M101VKY	Bova FLC12.280	Bova Futura Club	C53F	1995	

Previous Registrations:
GIL2785   F999JKR   JAZ9870   F22DUJ

Livery: Cream, brown and orange.

Pictured in the Harlescott area of Shrewsbury is Mercedes-Benz 811D F527BUX of Boultons. This vehicle   is fitted with Optare StarRider bodywork and features high-back seating. *Richard Godfrey*

5
5
Daily Mail
HYUNDAI
accent
BOWENS
...the first choice for holidays
BOWENS
E274 HRY

HOUSES
HOUSE
BOWENS
...the first choice for holidays
BOWENS
SCANIA
L405 LHE

BOWENS

L F Bowen Ltd, 101 Cotterills Lane, Birmingham B8 3SA

Depot : Fazeley Road, Tamworth.

D424POF	Bova FHD12.280	Bova Futura	C53F	1987	Ex Arnold, Tamworth, 1987
D230POF	Volvo B10M-61	Van Hool Alizée	C53F	1987	Ex Arnold, Tamworth, 1987
D784SGB	Volvo B10M-61	Plaxton Paramount 3500 III	C49FT	1987	Ex Park's, 1988
D785SGB	Volvo B10M-61	Plaxton Paramount 3500 III	C49FT	1987	Ex Park's, 1988
D786SGB	Volvo B10M-61	Plaxton Paramount 3500 III	C49FT	1987	Ex Park's, 1988
D787SGB	Volvo B10M-61	Plaxton Paramount 3500 III	C53F	1987	Ex Park's, 1988
E274HRY	Volvo B10M-61	Plaxton Paramount 3500 III	C51F	1988	Ex Smith's, Murton, 1990
E276HRY	Bova FHD12.290	Bova Futura	C49FT	1988	Ex Black Horse Travel, London, 1990
E673JNR	Bova FHD12.290	Bova Futura	C49FT	1988	Ex Black Horse Travel, London, 1990
E599UHS	Volvo B10M-61	Plaxton Paramount 3500 III	C49FT	1988	Ex Fords Travel, Gunnislake, 1990
F696ONR	Bova FHD12.290	Bova Futura	C49FT	1988	Ex Moseley demonstrator, 1989
F30COM	Bova FHD12.290	Bova Futura	C53F	1989	
F31COM	Bova FHD12.290	Bova Futura	C49FT	1989	
H619FUT	Bova FHD12.290	Bova Futura	C53F	1991	
H621FUT	Bova FHD12.290	Bova Futura	C49FT	1991	
H623FUT	Bova FHD12.290	Bova Futura	C53F	1991	
J405AWF	Bova FHD12.290	Bova Futura	C49FT	1992	
J407AWF	Bova FHD12.290	Bova Futura	C49FT	1992	
K713RNR	Toyota Coaster HDB30R	Caetano Optimo II	C18F	1992	
K714RNR	Toyota Coaster HDB30R	Caetano Optimo II	C18F	1992	
K296GDT	Bova FHD12.290	Bova Futura	C49FT	1993	
K297GDT	Bova FHD12.290	Bova Futura	C49FT	1993	
K298GDT	Bova FHD12.290	Bova Futura	C49FT	1993	
K299GDT	Bova FHD12.290	Bova Futura	C49FT	1993	
L405LHE	Scania K113CRB	Irizar Century 12.35	C49FT	1994	
L406LHE	Scania K113CRB	Irizar Century 12.35	C49FT	1994	
L407LHE	Scania K113CRB	Irizar Century 12.35	C49FT	1994	
L408LHE	Scania K113CRB	Irizar Century 12.35	C49FT	1994	
M316VET	Scania K113CRB	Irizar Century 12.35	C49FT	1995	
M317VET	Scania K113CRB	Irizar Century 12.35	C49FT	1995	
M318VET	Scania K113CRB	Irizar Century 12.35	C49FT	1995	
M319VET	Scania K113CRB	Irizar Century 12.35	C49FT	1995	
N	Scania K113CRB	Irizar Century 12.35	C49FT	1996	
N	Scania K113CRB	Irizar Century 12.35	C49FT	1996	
N	Scania K113CRB	Irizar Century 12.35	C49FT	1996	
N	Scania K113CRB	Irizar Century 12.35	C49FT	1996	
N	Scania K113CRB	Irizar Century 12.35	C49FT	1996	

Livery: Cream and red

Opposite, top: Bowens have become best known to the enthusiast for the contract at the National Exhibition Centre where they currently provide coaches between the car parks and halls. Seen on such a duty is E274HRY, a Volvo B10M with Plaxton Paramount 3500 bodywork. *Ralph Stephens*
Opposite, bottom: Since 1994 eight Scania K113s with Irizar Century bodies have been taken into stock with a further five expected as we go to press. One of the first quartet, L405LHE, is seen with the stars and stripes displayed while on touring duties. *Colin Lloyd*

BOYDON

DR DM RM & GM Boydon, Winkhill Filling Station, Ashbourne Road, Winkhill, Staffordshire.

w	AUP650L	Bedford YRQ		Plaxton Supreme III	C45F	1973	Ex Spratt, Wreningham, 1981
w	WWR424L	AEC Reliance 6U3ZR		Duple Dominant Express	C49F	1973	Ex Fisher-Ince, Market Raisen, 1982
w	NUS6P	AEC Reliance 6U3ZR		Plaxton Supreme III	C51F	1976	Ex Holmes, Moreton, 1989
w	KVE906P	AEC Reliance 6U3ZR		Plaxton Supreme III Express	C49F	1976	Ex Mather, Poulton-le-Fylde, 1995
	ADF871T	AEC Reliance 6U3ZR		Plaxton Supreme III Express	C53F	1978	Ex Mather, Poulton-le-Fylde, 1995
	JTU228T	Bedford YLQ		Plaxton Supreme IV	C45F	1979	Ex Bostocks, Congleton, 1996
	SIB7882	AEC Reliance 6U3ZR		Plaxton Supreme IV	C57F	1979	Ex George, Hare Street, 1988
	SIB3415	AEC Reliance 6U3ZR		Plaxton Supreme IV	C57F	1979	Ex Armstrong, Bletchley, 1988
	GSU7T	AEC Reliance 6U3ZR		Plaxton Supreme IV	C51F	1979	Ex Flear Coaches, Middlesbrough, 1985
	XBX831T	AEC Reliance 6U3ZR		Plaxton Supreme IV Express	C53F	1979	Ex Eynons, Trimsaran, 1986
	EBM449T	AEC Reliance 6U3ZR		Plaxton Supreme IV	C57F	1979	Ex Grange, East Ham, 1990
w	VVH861V	Ford Transit 190		Reeve Burgess	C17F	1979	Ex Moorland Rover, Werrington, 1991
w	CBK931W	Mercedes-Benz L508D		Robin Hood	C19F	1981	Ex Cooper, Elkstone, 1995
	SIB3053	AEC Reliance 6U3ZR		Plaxton Supreme IV	C53F	1980	Ex Dudley Coachways, 1993
	UIB4589	Bristol LHS6L		Plaxton Supreme IV	C35F	1981	Ex Courtesy, Werneth, 1994
	4195PX	Leyland Tiger TRCTL11/3R		Plaxton Paramount 3500	C49FT	1982	Ex Cadishead Coaches, 1993
	TIB2865	Leyland Tiger TRCTL11/3R		Plaxton Paramount 3200	C57F	1982	Ex Walls of Wigan, 1993
	MIB4964	Leyland Tiger TRCTL11/3R		Plaxton Paramount 3500	C50F	1983	Ex Bordian, Darwen, 1989
	VIB6165	Leyland Tiger TRCTL11/3R		Plaxton Paramount 3500	C53F	1983	Ex Gelsthorpe, Mansfield, 1995
	A891EBC	Mercedes-Benz L307D		Reeve Burgess	M12	1983	Ex Heritage Tours, Hyde, 1994
	NIW5986	Mercedes-Benz L608D		Reeve Burgess	C21F	1984	Ex Flintham, Metheringham, 1993
	RIB8034	Mercedes-Benz L608D		Mellor	C21F	1985	Ex Booth, Bury, 1992

Previous Registrations:

4195PX	WBV540Y	SIB3053	PRO445W	TIB2865	CLD886Y
MIB4964	THL293Y	SIB3415	EBM457T	UIB4589	SFP828X
NIW5986	A132MFL	SIB7882	EBM446T	VIB6165	BAJ632Y, 540CCY, SWN885Y
RIB8034	C612ADB				

Livery: White, red, yellow and orange.

Winkhill is located in the Leek to Ashbourne road and is home to the Boyden fleet of coaches many of which are found on school contracts for the county council. Seen prepared for service is VIB6165, a Leyland Tiger with Plaxton Paramount 3500 bodywork.
Bill Potter

BRITANNIA

Wrekin Coach Services Ltd, Travel House, 17 Market Street,
Oakengates, Telford TF2 6EL

Depot : Coach Garage, Donnington Wood, Telford.

1	NIW6131	DAF MB230DKFL615	Van Hool Alizée	C55F	1987	Ex Robinsons, Great Harwood, 1992
2	NIW2322	DAF MB230DKFL615	Van Hool Alizée	C55F	1987	Ex Robinsons, Great Harwood, 1992
3	GIL3273	DAF MB200DKFL600	Plaxton Paramount 3500 II	C53F	1985	Ex Happy Days, Woodseaves, 1987
4	GIL3274	DAF SB2300DHS585	Plaxton Paramount 3200	C55F	1984	Ex Meadway, Birmingham, 1988
5	GIL3275	Leyland Tiger TRCTL11/3R	Plaxton Paramount 3500 II	C53F	1984	Ex Robinson, Gr Harwood, 1995
6	NIW3546	Volvo B10M-60	Van Hool Alizée	C53F	1993	Ex Park's, 1995
7	NIW2317	DAF SB2305DHS585	Duple 340	C55F	1988	
8	GHE739V	Leyland Leopard PSU5C/4R	Plaxton Supreme IV	C57F	1980	Ex Hardman, Waterfoot, 1990
9	GIL1909	Leyland Tiger TRCTL11/3R	Plaxton Paramount 3200	C57F	1984	Ex Stevensons, 1991
10	LIL7810	Volvo B10M-61	Plaxton Paramount 3200 III	C53F	1987	Ex Express Travel, Perth, 1994
11	GNT434V	Ford R1114	Plaxton Supreme IV	C53F	1980	
12	LIL7812	DAF SB2305DHS585	Duple 340	C57F	1988	Ex Angel Coaches, Tottenham, 1996
14	LIL7814	Ford R1114	Plaxton Supreme IV	C53F	1980	
15	OIW7115	DAF MB230DKFL615	Van Hool Alizée	C55F	1987	Ex Fishwick, Leyland, 1994
16	TOF683S	Leyland National 11351A/1R		B52F	1978	Ex Midland, 1995
17	OIW5807	DAF SB2305DHTD585	Plaxton Paramount 3200 III	C57F	1988	Ex Midland Coaches, Auchterarder, 1994
18	NIW2318	DAF MB230DKFL615	Plaxton Paramount 3500 III	C53F	1987	

Further school work gained in 1994 prompted the introduction of double-deck operation in the form of a Bristol VRT, WTU485W. Seen in the Malinsgate district of Telford the vehicle is dedicated to school and works services. *Bill Potter*

The North & West Midlands Bus Handbook

Britannia Travel took 15, OIW7115 into stock in 1994. One of two DAF Van Hools previously with Fishwick at Leyland, they were painted from that scheme into the blue and yellow livery a few weeks after arrival. Seen in London on tour work it is one of a growing number of DAF-based coaches in the fleet. *Colin Lloyd*

19	WTU485W	Bristol VRT/SL3/6LXB	Eastern Coach Works	H43/31F	1981	Ex Midland, 1994
20	NIW2320	DAF MB230DKFL615	Van Hool Alizée	C55F	1987	Ex Fishwick, Leyland, 1994
21	NIW2321	Leyland Tiger TRCTL11/3RZ	Plaxton Paramount 3200 III	C57F	1987	Ex Priory Coaches, Gosport, 1990
22	KAD349V	Leyland Leopard PSU5C/4R	Plaxton Supreme IV	C57F	1980	Ex Cook, Biggleswade, 1986
23	OIW7023	DAF SB2300DHS585	Plaxton Paramount 3200	C53F	1984	Ex Wombwell Coaches, 1994
24	BGK314S	Leyland Leopard PSU5C/4R	Plaxton Supreme III	C55F	1978	Ex Epsom Coaches, 1987
25	GIL2195	Leyland Tiger TRCTL11/3R	Plaxton Paramount 3200	C57F	1983	Ex Warren, Tenterden, 1991
26	OIW7026	Leyland Tiger TRCTL11/3R	Plaxton Supreme V	C53F	1982	Ex Robinsons, Great Harwood, 1989
27	OIW7027	Leyland Tiger TRCTL11/3R	Plaxton Supreme V	C53F	1982	Ex Robinsons, Great Harwood, 1989
28	J129VAW	Mercedes-Benz 814D	Reeve Burgess Beaver	C33F	1991	
29	H177EJF	Toyota Coaster HB31R	Caetano Optimo	C21F	1991	
30	G453XHK	Mercedes-Benz 811D	Reeve Burgess Beaver	C25F	1989	Ex Cyril Evans, Senghenydd, 1995

Previous Registrations:

GHE739V	YHG189V, 170BHR	LIL7812	GNT433V	NIW3546	-
GIL1909	A834PPP	LIL7813	GNT432V	NIW6131	D218YCW
GIL2195	XCD138Y	NIW2317	E832YAW	OIW5807	E447YGG, 8733CD, E602LSL
GIL3273	B879AJX	NIW2318	E176VUJ	OIW7023	A828GFP
GIL3274	B996BOJ	NIW2320	D277XCX	OIW7026	LEC196X
GIL3275	-	NIW2321	D103ERU	OIW7027	LEC197X
LIL7810	D442CNR	NIW2322	D214YHG	OIW7115	D275XCX

Livery: Blue and yellow

BUTTERS

N S Butter & Partners, Yew Tree Cottage, Village Road, Childs Ercall, Market Drayton, Shropshire TF9 2DG

OOU534M	Bedford YRT	Plaxton Elite III	C53F	1973	Ex Epsom Coaches, 1979
TMJ633R	Bedford YMT	Duple Dominant II	C53F	1977	Ex Turner, Brown Edge, 1984
UWJ628S	Bedford YMT	Plaxton Supreme III	C53F	1978	Ex Shaw, Barnsley, 1981
UUX842S	Bedford YMT	Plaxton Supreme III Express	C53F	1978	Ex Williamsons, Shrewsbury, 1988
EAA830W	Bedford YLQ	Duple Dominant II	C45F	1980	Ex Walsall MBC, 1995
VWX352X	Bova EL26/581	Bova Europa	C51F	1982	Ex Kingswood Coaches, 1989
TFG221X	Leyland Leopard PSU5E/4R	Plaxton Supreme V	C50F	1982	Ex Brighton & Hove, 1989
HIL8863	DAF MB200DKFL600	LAG Galaxy	C53F	1983	Ex Zamir, Burton-on-Trent, 1993
AUT842Y	Mercedes-Benz L307D	Reeve Burgess	M12	1983	Ex Boulton, Cardington, 1986
DXI1454	Leyland Royal Tiger RTC	Leyland Doyen	C51F	1984	Ex Sandown, Padiham, 1992

Previous Registrations:
DXI1454   A483MHG   HIL8863   PBX280Y

Livery: White and blue

Set in rural countryside near Market Drayton the small Butter operation is now in the hands of the grandchildren of its founder. From the attractive village depot many services were introduced to support the military bases in the area at that time. Now most journeys are school and tendered services along with private hire. Seen in the village shortly after a repaint is TFG221X, a Leyland Leopard with Plaxton Supreme V bodywork. *Bill Potter*

The North & West Midlands Bus Handbook

CAVE

E H & G H Cave, 1-5 High Street, Solihull Lodge, Shirley, Solihull, West Midlands   B90 1HA

	HHA126L	Leyland National 1151/1R/2501		B51F	1973	Ex Kirby, Wythall, 1991
w	GBF73N	Leyland National 11351/1R		B49F	1975	Ex PMT, 1992
	HNB24N	Leyland National 10351/1R		B41F	1975	Ex Cambrian Coast, Abergynolwyn, 1994
	LIL3059	Leyland National 11351/1R		B49F	1975	Ex Victoria Travel, Earlestown, 1995
w	KDW360P	Leyland National 11351/1R/SC		DP48F	1975	Ex Red & White, 1994
	NWO476R	Leyland National 11351A/1R		B52F	1976	Ex Red & White, 1994
	NWO483R	Leyland National 11351A/1R		B52F	1977	Ex Red & White, 1994
	NWO493R	Leyland National 11351A/1R		B52F	1976	Ex Red & White, 1994
	NWO499R	Leyland National 11351A/1R		B52F	1977	Ex Red & White, 1994
	TUP432R	Bedford YMT	Plaxton Derwent	DP53F	1976	Ex Armstrong, Ebchester, 1983
	TTC534T	Leyland National 11351A/1R		B52F	1978	Ex Metrowest, Coseley, 1993
	YPL397T	Leyland National 10351B/1R		B41F	1978	Ex D&G, Rachub, 1993
	KRP562V	Leyland National 11351A/1R		B52F	1979	Ex Thornton, Messing, 1995
	J921TUK	ACE Cougar	Willowbrook Warrior	B40F	1991	

Previous Registrations:
LIL3059   KRE278P

Livery: Turquoise and grey

Cave's fleet is dominated by the Leyland National and an ACE Cougar on services in the borough of Solihull, part of the West Midlands county. Seen at the town's interchange is YPL397T. *Bill Potter*

CHASE BUS SERVICES

Chase Coaches Ltd, No Name Road, Chasetown, Walsall, Staffordshire   WS7 8FS

1	THX117S	Leyland National 10351A/2R		B41F	1978	Ex London Buses, 1992
2	OJD858R	Leyland National 10351A/2R		B44F	1977	Ex London Buses, 1992
3	AYR309T	Leyland National 10351A/2R		B35D	1979	Ex Eastbourne, 1990
4	YYE274T	Leyland National 10351A/2R		B35D	1979	Ex Eastbourne, 1990
5	AYR339T	Leyland National 10351A/2R		B35D	1979	Ex Eastbourne, 1990
6	YYE295T	Leyland National 10351A/2R		B35D	1979	Ex Eastbourne, 1990
7	THX151S	Leyland National 10351A/2R		B36D	1978	Ex London Buses, 1990
8	THX181S	Leyland National 10351A/2R(0680)		B42F	1978	Ex London Buses, 1990
9	BYW365V	Leyland National 10351A/2R		B36D	1979	Ex London Buses, 1990
10	MRP5V	Leyland National 10351A/2R(0680)		B43F	1979	Ex London Buses, 1990
11	THX260S	Leyland National 10351A/2R		B36D	1978	Ex London Buses, 1990
12	OJD863R	Leyland National 10351A/2R		B36D	1977	Ex London Buses, 1990
14	MMB970P	Leyland National 11351/1R/SC		DP48F	1976	Ex Crosville Wales, 1990
15	THX222S	Leyland National 10351A/2R		B36D	1978	Ex London Buses, 1990
16	THX236S	Leyland National 10351A/2R		B40D	1978	Ex London Buses, 1990
17	DDW429V	Leyland National 10351A/2R		B41F	1979	Ex The Wright Company, Wrexham, 1994
18	HHU632N	Leyland National 10351/1R		B44F	1975	Ex County, Leicester, 1989
19	NEN957R	Leyland National 11351A/1R		B49F	1977	Ex County, Leicester, 1989
21	NEN965R	Leyland National 11351A/1R		B49F	1977	Ex County, Leicester, 1989
22	NWO467R	Leyland National 11351A/1R/SC		DP48F	1977	Ex Burrows, Ogmore Vale, 1993
23	HUH409N	Leyland National 10351/1R		B41F	1975	Ex Amberley, Pudsey, 1989
24	WBN464T	Leyland National 11351A/1R		B49F	1979	Ex Tanat Valley, Pentrefelin, 1993
25	LTG796P	Leyland Leopard PSU3C/2R	East Lancashire	B51F	1976	Ex Inter Valley Link, 1989
26	LTG797P	Leyland Leopard PSU3C/2R	East Lancashire	B51F	1976	Ex Inter Valley Link, 1989
27	LTG798P	Leyland Leopard PSU3C/2R	East Lancashire	B51F	1976	Ex Inter Valley Link, 1989
28	LTG850P	Leyland Leopard PSU4C/2R	East Lancashire	B45F	1976	Ex Inter Valley Link, 1989
29	PPM894R	Leyland National 11351A/1R		B49F	1977	Ex Clyde Coast, Ardrossan, 1993
30	YBJ403	DAF SB220LC550	Ikarus Citibus	B49F	1995	Ex Delta, Mansfield, 1996
31	THX160S	Leyland National 10351A/2R		B36D	1978	Ex London Buses, 1992
32	RKA870T	Leyland National 11351A/1R		B52F	1979	Ex Merseybus, 1991
33	BYW382V	Leyland National 10351A/2R(Volvo)		B44F	1979	Ex Parfitt's Rhymney Bridge, 1995
34	KNW656N	Leyland National 11351/1R		B52F	1975	Ex West Yorkshire, 1988
35	BYW357V	Leyland National 10351A/2R(Volvo)		B44F	1979	Ex Parfitt's Rhymney Bridge, 1995
36	HCA969N	Leyland National 11351/1R/SC		B48F	1974	Ex Crosville Wales, 1987
38	HMA658N	Leyland National 11351/1R/SC		B48F	1975	Ex Crosville, 1987
39	HFM183N	Leyland National 11351/1R/SC		B48F	1974	Ex Crosville Wales, 1987
40	THX209S	Leyland National 10351A/2R		B36D	1978	Ex London Buses, 1990
41	EGB78T	Leyland National 11351A/1R		B52F	1979	Ex Western Scottish, 1991
42	THX193S	Leyland National 10351A/2R		B36D	1978	Ex London Buses, 1990
43	OJD868R	Leyland National 10351A/2R		B35D	1977	Ex Eastbourne, 1991
44	OJD870R	Leyland National 10351A/2R		B36D	1977	Ex Black Prince, Morley, 1990
45	EGB94T	Leyland National 11351A/1R		B52F	1979	Ex Western Scottish, 1991

Lichfield Street, Wolverhampton is the location of this picture of Chase 26, LTG797P, a Leyland Leopard with East Lancashire bodywork previously with the Inter Valley Link operation of Rhymney Valley in south Wales.
Richard Godfrey

CHASE
VANHOOL
PSU 977
CHASE

VIA SCOTT ARMS
50
WALSALL
VIA SCOTT ARMS
CHASE BUS SERVICES
CHASE
CHASE
35
BYW 357V

46	OJD865R	Leyland National 10351A/2R		B36D	1977	Ex London Buses, 1992
47	RTE111G	Leyland Leopard PSU3A/2R	East Lancashire	B51F	1969	Ex Lancaster, 1983
48	TEL493R	Leyland National 11351A/2R		B48F	1977	Ex Wilts & Dorset, 1994
49	BYW369V	Leyland National 10351A/2R(Volvo)		B44F	1979	Ex Parfitt's Rhymney Bridge, 1995
50	THX149S	Leyland National 10351A/2R		B44F	1978	Ex Parfitts, Rhymney Bridge, 1993
51	THX159S	Leyland National 10351A/2R		B44F	1978	Ex Parfitts, Rhymney Bridge, 1993
52	THX266S	Leyland National 10351A/2R		B44F	1978	Ex Parfitts, Rhymney Bridge, 1993
53	THX264S	Leyland National 10351A/2R		B44F	1978	Ex Parfitts, Rhymney Bridge, 1993
54	YYE270T	Leyland National 10351A/2R		B36D	1979	Ex Parfitts, Rhymney Bridge, 1993
55	AYR317T	Leyland National 10351A/2R		B44F	1979	Ex Parfitts, Rhymney Bridge, 1993
56	AYR330T	Leyland National 10351A/2R		B44F	1979	Ex Parfitts, Rhymney Bridge, 1993
57	AYR343T	Leyland National 10351A/2R		B44F	1979	Ex Parfitts, Rhymney Bridge, 1993
58	YPF773T	Leyland National 10351A/1R		B41F	1979	Ex Sovereign, 1993
59	SPC287R	Leyland National 10351A/1R		B41F	1979	Ex Sovereign, 1993
60	D866NVS	Freight Rover Sherpa	Dormobile	B16F	1986	Ex Humphries, Pelsall, 1994
61	BYW358V	Leyland National 10351A/2R(Volvo)		B44F	1979	Ex Parfitt's Rhymney Bridge, 1995
62	BYW366V	Leyland National 10351A/2R(Volvo)		B44F	1979	Ex Parfitt's Rhymney Bridge, 1995
63	TRN809V	Leyland National 10351B/1R		B44F	1979	Ex Ribble, 1995
64	TRN808V	Leyland National 10351B/1R		B44F	1979	Ex Ribble, 1995
65	TRN807V	Leyland National 10351B/1R		B44F	1979	Ex Ribble, 1995
250	E834EVS	Ford Transit VE6		M16	1988	
251	PSU988	Volvo B10M-61	Caetano Algarve	C53F	1986	Ex Park's, 1990
252	PSU989	Volvo B10M-61	Caetano Algarve	C53F	1986	Ex Price, Halesowen, 1990
254	FSV428	Volvo B10M-60	Plaxton Paramount 3500 III	C49FT	1983	Ex Harry Shaw, 1992
255	PSU954	Leyland Tiger TRCTL11/3R	Plaxton Viewmaster IV	C50F	1983	Ex Horlock, Northfleet, 1988
256	PSU942	Volvo B10M-61	Van Hool Alizée	C49FT	1988	Ex Excelsior, 1993
257	PSU969	Volvo B10M-61	Van Hool Alizée	C49FT	1984	Ex Shearings, 1991
259	PSU987	DAF MB230DKFL615	Jonckheere Jubilee P50	C51F	1988	Ex Hallmark, Luton, 1992
260	PSU946	Volvo B10M-61	Plaxton Paramount 3500 III	C49FT	1988	Ex Limebourne, Battersea, 1993
262	E225WWD	Ford Transit VE6	Dormobile	M16	1988	Ex Driver, Rainham, 1993
263	G880ELJ	Ford Transit VE6	Zodiac	M14	1990	Ex Conveyor Services Europe, 1993
264	PSU906	Volvo B10M-61	Plaxton Paramount 3500 III	C49FT	1988	Ex West Coast Motors, Campbeltown, 1994
265	PSU977	DAF MB230LT615	Van Hool Alizée	C53FT	1989	Ex Welsh's, Pontefract, 1994
266	G929GWN	Ford Transit VE6	Ford	M12	1990	Ex Castle Garage, Llandovery, 1994

Previous Registrations:

E225WWD	E860WYC, PSU987, YBJ403	PSU942	XEL254, E749SEL	PSU987	E682NNH, YBJ403
FSV428	F32VAC, COV8V, KJI3935	PSU946	E302OMG	PSU988	C678KDS
KBC110N	JND998N, 531PP	PSU954	KGS488Y	PSU989	C682KDS
MRP5V	BYW418V	PSU969	A197MNE	YBJ403	M835RCP
PSU906	E360XSB	PSU977	F655OHD		

Livery: Orange, brown and white; Coaches - white, red and blue or Silver blue metallic.

Opposite: **A contrast between the bus and coach operations is seen here with Chase 265, PSU977 and Leyland National 35, BYW357V. Metallic blue is used for the coach livery and 265, seen in London, carries the stylised C and stagshead a reference to the former royal hunting grounds of Cannock Chase. Leyland National 35 was new to London as dual-door though, as can be seen, now has a single entrance/exit.** *Richard Godfrey*

Chase number the coach fleet in a series above 250. Pictured passing through Parliament Square is 254, FSV428, a vehicle previously with Harry Shaw. 1996 is the companys 30th anniversary - a DAF SB220 being numbered 30 and lettered accordingly has just entered service.
Colin Lloyd

23

CHOICE TRAVEL

Midland Choice Travel Ltd; Liyell Ltd, Watery Lane, Willenhall, West Midlands   WV13 3SU

1	TUB7M	Leyland Leopard PSU4B/4R	Plaxton Supreme III	C43F	1974	Ex Richardson, Midhurst, 1992
3	EGB81T	Leyland National 11351A/1R		B49F	1978	Ex Western Scottish, 1992
5	ERP559T	Leyland National 11351A/1R		B49F	1979	Ex United Counties, 1992
6	SBD525R	Leyland National 11351A/1R		B49F	1977	Ex United Counties, 1992
8	PCD74R	Leyland National 11351A/1R		B49F	1976	Ex Stagecoach South, 1992
9	SFJ139R	Leyland National 11351A/1R		B49F	1977	Ex Stagecoach South, 1992
10	XLD627	Leyland TRCTL11/3RH	Plaxton Paramount 3500	C51F	1984	Ex Tayside, 1993
12	D36KAX	Iveco Daily 49.10	Robin Hood City Nippy	B21F	1986	Ex Rhondda, 1992
14	PTF751L	Leyland National 1151/2R/0402		B49F	1973	Ex Ribble, 1993
15	MLG961P	Leyland National 11351/1R/SC		B52F	1975	Ex Hopkinson, Market Harborough, 1993
16	GMA409N	Leyland National 11351/1R/SC		B46F	1974	Ex Ogden's, Haydock, 1993
19	WOC731T	Leyland Leopard PSU3E/4R	Plaxton Supreme III Express	C49F	1978	Ex Midland Red West, 1994
20	A122BHL	Leyland Royal Tiger B50	Plaxton Paramount 3500	C50F	1984	Ex Yorkshire Traction, 1994
21	TJN509R	Leyland National 11351A/1R		B49F	1977	Ex Thamesway, 1994
22	JTH782P	Leyland National 11351A/1R		B52F	1976	Ex Thamesway, 1994
23	JTH783P	Leyland National 11351A/1R		B52F	1976	Ex Thamesway, 1994
24	PJT256R	Leyland National 10351A/1R		B44F	1976	Ex Thamesway, 1994
25	G149GOL	Iveco Daily 49.10	Carlyle Dailybus 2	B25F	1990	Ex Moffat & Williamson, Gauldry, 1995
26	M26XEH	Dennis Dart 9.8SDL3054	Northern Counties Paladin	B39F	1995	
27	M7TUB	Dennis Dart 9.8SDL3054	Northern Counties Paladin	B39F	1995	
28	JOX530P	Leyland National 11351A/1R		B49F	1976	Ex Midland, 1995
29	TOF691S	Leyland National 11351A/1R		B49F	1978	Ex Midland, 1995
30						
31	N31EVT	Mercedes-Benz 1416	Wright Urbanranger	B47F	1995	
32	N32EVT	Mercedes-Benz 1416	Wright Urbanranger	B47F	1995	
33	N133GRF	Mercedes-Benz 1416	Wright Urbanranger	B47F	1996	

Previous Registrations:
A122BHL   A73YDT, 5562HE   XLD627   A710SSR

Livery: Green and yellow

Choice Travel's fleet is represented by three types of single-deck bus. *Right* is National JTH782P new to South Wales Transport is seen here in Walsall. *Opposite,* are Northern Counties-bodied Dart M26XEH and N32EVT, a Mercedes-Benz 1416 with Wright UrbanRanger body. Fleet number 30 will be used by a 1996 delivery.
Cliff Beeton/Richard Godfrey/Tony Wilson

Bloxwich
MOSSLEY
171
Bank
Choice
TRAVEL
DENNIS
Choice
M26 XEH

LOCK WORKS
WOLVERHAMPTON
510
Choice
TRAVEL
Choice
N32 EVT

Claribels operate into the centre of Birmingham. Contrasting the NBC express coach deliveries of the late 1970s and early 1980s are NEL111P with Plaxton Supreme bodywork and WCK132V with Duple Dominant II styles new to Alder Valley and Ribble respectively. *Tony Wilson/Paul Wigan*

CLARIBELS

Claribel (APH) Coaches Ltd, 10 Fortnum Close, Tile Cross, Birmingham B33 0JT

	JOX458P	Leyland Leopard PSU3C/4R	Plaxton Supreme III Express	C49F	1976	Ex Midland Red West, 1993
	NEL111P	Leyland Leopard PSU3C/4R	Plaxton Supreme III Express	C49F	1976	Ex Andy James, Tetbury, 1994
w	MAW112P	Ford R1114	Plaxton Supreme III	C49F	1976	Ex Allenways, Birmingham, 1991
	PHH613R	Leyland Leopard PSU3C/4R	Duple Dominant Express	C49F	1978	Ex Midland Red South, 1993
	SOA673S	Leyland Leopard PSU3E/4R	Plaxton Supreme III Express	C49F	1977	Ex Midland Red West, 1994
	BGY589T	Leyland Leopard PSU5C/4R	Plaxton Supreme IV	C48FT	1979	Ex Yardleys, Birmingham, 1993
	JMB330T	Leyland Leopard PSU3E/4R	Duple Dominant I	C53F	1979	Ex Stephenson, Rochford, 1993
	JMB337T	Leyland Leopard PSU3E/4R	Duple Dominant I	C53F	1979	Ex Stephenson, Rochford, 1993
	KRN110T	Leyland Leopard PSU3E/4R	Duple Dominant II Express	C53F	1979	Ex Birmingham Omnibus, Tividale, 1995
	KRN115T	Leyland Leopard PSU3E/4R	Duple Dominant II Express	C53F	1979	Ex Birmingham Omnibus, Tividale, 1995
	ODM501U	Leyland Leopard PSU3E/4R	Duple Dominant II Express	C53F	1979	Ex Stephenson, Rochford, 1993
	KRO645V	Ford R1114	Duple Dominant II	C45F	1980	Ex Golden Boy, Hoddesdon, 1986
	WCK132V	Leyland Leopard PSU3E/4R	Duple Dominant II Express	C49F	1979	Ex Midland Red North, 1993
	FYX812W	Leyland Leopard PSU3E/4R	Duple Dominant II Express	C49F	1980	Ex Phil Anslow, Garndiffaith, 1994
	FYX813W	Leyland Leopard PSU3E/4R	Duple Dominant II Express	C53F	1980	Ex Phil Anslow, Garndiffaith, 1994
	FYX823W	Leyland Leopard PSU3E/4R	Duple Dominant II Express	C53F	1980	Ex Phil Anslow, Garndiffaith, 1994
	WWA299Y	Leyland Tiger TRCTL11/3R	Plaxton Paramount 3500	C51F	1983	Ex Allenways, Birmingham, 1992
	D900MWR	Freight Rover Sherpa	Dormobile	B20F	1987	Ex Yorkshire Rider, 1991
	CAZ2747	Leyland Royal Tiger RTC	Leyland Doyen	C49FT	1988	Ex Allenways, Birmingham, 1993
	CAZ2748	Leyland Royal Tiger RTC	Leyland Doyen	C49FT	1988	Ex Allenways, Birmingham, 1993
	E637KCX	DAF SB2305DHS585	Duple 340	C53F	1988	Ex Angel Motors, Tottenham, 1994
	CAZ2749	Hestair-Duple 425 SDA1512	Duple 425	C53FT	1989	Ex Allenways, Birmingham, 1993
	F472RPG	Renault Trafic	Jubilee	M11	1989	Ex van, 1990
	G368MFD	Ford Transit VE6	Jubilee	M14	1990	Ex van, 1990
	G155XJF	Toyota Coaster HB31R	Caetano Optimo	C21F	1990	Ex Allenways, Birmingham, 1992
	M944XET	Bova FHD12.340	Bova Futura	C51FT	1995	

Livery: White and blue

Previous Registrations:
CAZ2747	E53TYG		CAZ2748	E52TYG	CAZ2749	F32KHS

CLOWES

GA & KM Clowes & M Barks, Barrow Moor Farm, Barrow Moor, Longnor, Staffordshire SK17 0QP

MRH162P	Bedford YRQ	Duple Dominant Express	C45F	1976	Ex Leech, Macclesfield, 1988
RAW32R	Bedford YMT	Duple Dominant	C53F	1977	Ex Bryant's Coaches, Williton, 1989
CYH578V	Ford Transit 160	Clowes	M12	1980	Ex private owner, 1987
VRY1X	DAF MB200DKTL600	Smit Euro Hi-Liner	C53F	1982	Ex Robin Hood, Rushton, 1993
YXI6366	DAF MB200DKTL600	Plaxton Supreme IV	C53F	1981	Ex Happy Days, Woodseaves, 1995
YXI6367	Volvo B10M-56	Plaxton Supreme IV Express	C45F	1982	Ex McCarthy & Lomas, Macclesfield, 1994
YXI6246	Volvo B10M-61	Duple Goldliner	C53F	1982	Ex McGinley, Falcarragh, 1994
YXI7340	Bedford YNV Venturer	Willowbrook Crusader	C51FT	1986	Ex Alpha Coaches, Coventry, 1992
C777KGB	Freight Rover Sherpa	Scott	M16	1986	Ex Wint, Butterton, 1994
D770JUB	Freight Rover Sherpa	Dormobile	B20F	1986	Ex Yorkshire Rider, 1991
D710SKU	Freight Rover Sherpa	Crystals	M16	1986	Ex private owner, 1988
RIB3524	Aüwaerter Neoplan N216H	Aüwaerter Jetliner	C53F	1986	Ex Robin Hood, Rudyard, 1993

Previous Registrations:
RIB3524	C718JTL	YXI6366	GTC820X	YXI7340	C386VBC
YXI6246	FHS739X, 82DL264	YXI6367	KNP2X		

Livery: Orange and green.

COPELAND'S

Copeland Tours (Stoke on Trent) Ltd, Uttoxeter Road, Meir, Stoke-on-Trent ST1 3ER

MIB520	Ford R1114	Plaxton Supreme III	C43DL	1977	
RWM582T	Leyland Leopard PSU5C/4R	Plaxton Supreme IV	C53FT	1979	Ex Poole, Chesterton, 1995
MIB864	DAF MB200DKTL600	Jonckheere Bermuda	C55F	1982	Ex Slatepearl, Trentham, 1985
CVN347Y	DAF MB200DKTL600	Jonckheere Bermuda	C49FT	1982	Ex J&C, Newton Aycliffe, 1996
MIB614	Leyland Tiger TRCTL11/3R	Plaxton Paramount 3200 E	C50FT	1983	Ex Wessex, Bristol, 1991
MIB104	DAF MB200DKFL600	Plaxton Paramount 3200	C48FT	1983	Ex Bruce, Pitscottie, 1994
MIB268	DAF MB200DKFL600	Van Hool Alizée	C50FT	1983	Ex Smithson, Spixworth, 1996
MIB615	DAF MB200DKFL600	Plaxton Paramount 3200	C42FTL	1984	Ex Kinch, Barrow-on-Soar, 1988
MIB246	Leyland Tiger TRCTL11/3R	Plaxton Paramount 3500	C53F	1985	
MIB516	DAF MB200DKFL600	Jonckheere Jubilee P50	C51FT	1986	Ex Hardings, Redditch, 1994
MIB116	DAF MB200DKFL600	Jonckheere Jubilee P50	C51FT	1986	Ex Hardings, Redditch, 1994
D156LTA	Renault-Dodge S56	Reeve Burgess	B23F	1986	Ex Bridge Cs, Paisley, 1994
D164LTA	Renault-Dodge S56	Reeve Burgess	B23F	1986	Ex Cardiff Bus, 1994
MIB279	DAF MB230DKFL615	Plaxton Paramount 3500 III	C53F	1987	Ex Stevensons, 1994
MIB970	Mercedes-Benz 811D	North West Coach Sales	C21F	1987	Ex C&M, Aintree, 1989
MIB236	Leyland Tiger TRCL10/3ARZM	Plaxton Paramount 3500 III	C53FT	1988	Ex Volvo demonstrator, 1990
MIB761	Toyota Coaster HB31R	Caetano Optimo	C21F	1989	Ex Kinch, Barrow-on-Soar, 1993
MIB346	Hestair Duple SDA1512	Duple 425	C53FT	1989	Ex Grimshaw, Burnley, 1994

Previous Registrations:
CVN347Y	914PP, 205PP, DJF994Y, 205PP, DFP239Y, 71XVO		
MIB104	TTG244Y	MIB516	C910UUY, DSK566, C101VUY
MIB116	C909UUY, DSK565, C102VUY	MIB520	TRE202R, 111WEH
MIB236	F683SRN	MIB536	-
MIB246	B888SEH	MIB537	-
MIB268	KYC729	MIB614	EAH892Y, CIW6752, FFA270Y
MIB270	-	MIB615	A102HJF
MIB278	-	MIB746	-
MIB279	D606YCX, AAX568A, LUY742, D138DFP	MIB761	G860WBC
MIB302	-	MIB864	YRF754Y, 470WYA
MIB346	F545YCK	MIB905	C804CBU
MIB394	-	MIB970	D840OVM

Livery: Blue, navy and orange

Copeland's have built a collection of MIB index marks which are often retained dormant when vehicles are sold. One of the latest swaps is MIB516, a DAF MB230 with Jonckheere Jubilee bodywork seen at the base near Stoke-on-Trent. *Bill Potter*

ELCOCK REISEN

M H Elcock & Son Ltd, The Maddocks, Madeley, Telford, Shropshire TF7 5HA

Depots : The Maddocks, Madeley, Telford and Admaston Road, Wellington, Telford

WAW356S	Ford R1114	Plaxton Supreme III	C53F	1978	Ex Shropshire Education, 1990
ATH4V	Ford R1114	Plaxton Supreme IV	C53F	1979	Ex Castle Garage, Llandovery, 1989
HUJ998V	Ford R1114	Plaxton Supreme IV	C53F	1980	
HUJ999V	Ford R1114	Plaxton Supreme IV	C53F	1980	
JAW84V	Ford R1114	Plaxton Supreme IV	C53F	1980	Ex Excelsior, Telford, 1984
XDF7X	Ford R1114	Duple Dominant IV	C53F	1982	Ex Shropshire Education, 1994
1577NT	Volvo B10M-61	Plaxton Paramount 3500 II	C53F	1986	Ex Clarkes of London, 1992
E222WUX	Volvo B10M-61	Plaxton Paramount 3200 III	C53F	1987	
E261WWD	Leyland Tiger TRCTL11/3R	Plaxton Derwent II	DP53F	1987	
E961YUX	Ford Transit VE6	Ford	M15	1988	Ex MoD, 1996
E666YAW	Mercedes-Benz 609D	Reeve Burgess	C25F	1988	Ex Ford, Warley, 1990
3408NT	Volvo B10M-61	Plaxton Paramount 3500 III	C49FT	1988	
F482WFX	Mercedes-Benz 811D	Reeve Burgess Beaver	C29F	1989	Ex Excelsior, 1992

Elcock operate several tours to Bavaria, Austria and Switzerland, and have taken the German for tours - Reisen - into their name. Vehicles also feature a representation of the Telford Iron Bridge, though the depot in the Ironbridge district ceased to be used several years ago, the main depot now being at Wellington, with the engineering base at Madeley. Seen in London is HIL 6584, a 1989 Volvo B10M with Plaxton Paramount 3500 bodywork. *Colin Lloyd*

Local tendered services also feature in the Elcock portfolio. Seen leaving Wellington bus station is PUJ925 a Mercedes-Benz 814 with Plaxton Beaver bodywork. Several new minibuses have been added to the fleet, these often to be found on stage and contract work in the area. The Fords are dedicated to contract work that includes school duties. *Bill Potter*

EIL829	Volvo B10M-60	Plaxton Paramount 3200 III	C53FT	1989	Ex Excelsior, 1993
EIL1607	Volvo B10M-60	Plaxton Paramount 3200 III	C53FT	1989	Ex Excelsior, 1993
5038NT	Volvo B10M-60	Plaxton Paramount 3500 III	C53FT	1989	Ex Park's, 1991
HIL6584	Volvo B10M-60	Plaxton Paramount 3500 III	C53FT	1989	Ex Park's, 1991
G333JUX	Volvo B10M-61	Plaxton Paramount 3500 III	C49FT	1989	
G444JAW	Mercedes-Benz 811D	Reeve Burgess Beaver	C25F	1989	
1398NT	Volvo B10M-60	Plaxton Paramount 3500 III	C53F	1990	Ex Park's, 1991
EIL2247	Volvo B10M-60	Plaxton Paramount 3500 III	C49FT	1990	Ex Park's, 1991
PUJ925	Mercedes-Benz 811D	Reeve Burgess Beaver	C33F	1990	Ex Browns, Kirkby, 1996
K337ABH	Mercedes-Benz 814D	Plaxton Beaver	C33F	1992	Ex Brents Travel, 1996
K321AUX	Volvo B10M-60	Jonckheere Deauville P599	C51FT	1992	
K123CAW	Volvo B10M-60	Jonckheere Deauville P599	C51FT	1993	
K882DUJ	Mercedes-Benz 814D	Plaxton Beaver	C33F	1993	
L321JUJ	Volvo B10M-62	Plaxton Première 350	C49FT	1994	
M123OUX	Mercedes-Benz 814D	Plaxton Beaver	C33F	1994	
M123RAW	Volvo B10M-62	Plaxton Première 350	C49FT	1995	
M321RAW	Volvo B10M-62	Plaxton Première 350	C49FT	1995	

Previous Registrations:

1398NT	G59RGG	EIL1607	F462WFX, XEL158, F769MAA
1577NT	C177LWB	EIL2247	G87RGG
3408NT	F555CAW	EIL829	F461WFX, XEL31, XEL24, F755MAA
5038NT	F985HGE	HIL6584	F966HGE
E961YUX	E......, 1398NT	PUJ925	H155WRB

Livery: Silver, red and gold.

FALCON TRAVEL

Falcon Travel Ltd, Unit 14, London Street Trading Estate, Smethwick,
West Midlands B66 2QZ

	WNO558L	Leyland National 1151/1R/0401	B52F	1973	Ex Baldwin, Sheffield, 1994
	HHA115L	Leyland National 1151/1R/2501	B52F	1973	Ex C&G, Chatteris, 1984
	XRR584M	Leyland National 1151/2R/0403	B49F	1973	Ex East Midland, 1994
	OAO564M	Leyland National 1151/1R/0401	B52F	1974	Ex Bajwa, Slough, 1995
	NFN69M	Leyland National 1151/1R/2402	B49F	1974	Ex Bajwa, Slough, 1995
	GLJ680N	Leyland National 11351/1R	DP48F	1975	Ex Victoria Travel, Earlestown, 1994
	SGR565R	Leyland National 11351A/1R	B49F	1976	Ex Waddell, Lochwinnoch, 1993
w	NEN958R	Leyland National 11351A/1R	B49F	1977	Ex Thanet Bus, Ramsgate, 1993
	MDL881R	Leyland National 11351A/1R	B52F	1978	Ex ABC Travel, Olney, 1994
	VFX983S	Leyland National 11351A/1R	B49F	1978	Ex Wilts & Dorset, 1993
	XNG773S	Leyland National 11351A/1R	B52F	1978	Ex Thanet Bus, Ramsgate, 1993
	BAL608T	Leyland National 11351A/1R	B52F	1978	Ex MTL, 1995
	WBN478T	Leyland National 11351A/1R	B49F	1979	Ex Thanet Bus, Ramsgate, 1993
	AYR324T	Leyland National 10351A/2R	B36D	1979	Ex Bygone Buses, Biddenden, 1994
	JHJ141V	Leyland National 11351A/1R	B49F	1979	Ex MTL, 1995

Livery: Maroon and Cream

Pictured in Carrs Lane, Birmingham while heading for Druids Heath is Falcon Travel's National MDL881R once new to Southern Vectis. Southern national only took two batches of new 11-metre Nationals, five in 1973 and three in 1976, and as such was probably one of NBC fleets that used the type the least. A maroon and cream livery is carried by this all-Leyland National fleet. *Tony Wilson*

FLIGHTLINK
Manchester
Birmingham
Heathrow
Gatwick
FLIGHTLINK
HEATHROW
GATWICK
L44 FTG
local

LOCK WORKS
FLIGHTS
FLIGHTS
H2 FTG
WARNING!

FLIGHTS

Flights Coach Travel Ltd, Beacon House, Long Acre, Nechells,
Birmingham B7 5JJ

TOB377	AEC Reliance MU3RV	Burlingham Seagull	C37C	1956	Ex preservation, 1995
1FTG	Bova FHD12.280	Bova Futura	C25FT	1989	Ex Central Coachways, 1993
A2FTG	Volvo B10M-50	Plaxton Paramount 4000 III	CH55/12DT	1990	
A3FTG	Volvo B10M-50	Plaxton Paramount 4000 III	CH55/12DT	1990	
A4FTG	Volvo B10M-50	Plaxton Paramount 4000 III	CH55/12DT	1990	
H2FTG	Toyota Coaster HDB30R	Caetano Optimo II	C18F	1991	
H3FTG	Toyota Coaster HDB30R	Caetano Optimo II	C18F	1991	
J4FTG	Toyota Coaster HDB30R	Caetano Optimo II	C19F	1992	
J6FTG	Toyota Coaster HDB30R	Caetano Optimo II	C21F	1992	
K12FTG	Volvo B10M-60	Plaxton Excalibur	C49FT	1992	
K16FTG	Volvo B10M-60	Plaxton Excalibur	C49FT	1993	
K17FTG	Volvo B10M-60	Plaxton Excalibur	C49FT	1993	
K18FTG	Volvo B10M-60	Plaxton Excalibur	C49FT	1993	
K19FTG	Volvo B10M-60	Plaxton Excalibur	C49FT	1993	
K20FTG	Volvo B10M-60	Plaxton Excalibur	C49FT	1993	
L22FTG	Bova FHD12.340	Bova Futura	C44FT	1994	
L33FTG	Bova FHD12.340	Bova Futura	C44FT	1994	
L44FTG	Bova FHD12.340	Bova Futura	C44FT	1994	
L55FTG	Bova FHD12.340	Bova Futura	C44FT	1994	
L66FTG	Bova FHD12.340	Bova Futura	C44FT	1994	
L77FTG	Bova FHD12.340	Bova Futura	C44FT	1994	
L1NER	Bova FHD12.340	Bova Futura	C44FT	1994	
L1NKF	Bova FHD12.340	Bova Futura	C44FT	1994	
M1FTG	Aüwaerter Neoplan N122/3	Aüwaerter Skyliner	CH57/12CT	1995	
M2FTG	Aüwaerter Neoplan N122/3	Aüwaerter Skyliner	CH57/12CT	1995	
M3FTG	Aüwaerter Neoplan N122/3	Aüwaerter Skyliner	CH57/12CT	1995	
M10FTG	Bova FHD12.340	Bova Futura	C44FT	1995	
M20FTG	Bova FHD12.340	Bova Futura	C44FT	1995	
M30FTG	Bova FHD12.340	Bova Futura	C44FT	1995	
M40FTG	Bova FHD12.340	Bova Futura	C44FT	1995	
M50FTG	Bova FHD12.340	Bova Futura	C44FT	1995	
M60FTG	Bova FHD12.340	Bova Futura	C44FT	1995	
N495TVP	Nissan Serena	Nissan	M7	1995	
2FTG	Volvo B10M-62	Plaxton Excalibur	C46FT	1996	
FTG5	Volvo B10M-62	Plaxton Excalibur	C46FT	1996	
FTG9	Volvo B10M-62	Plaxton Excalibur	C46FT	1996	
	Volvo B10M-SE	Plaxton Excalibur	C49FT	1996	
	Volvo B10M-SE	Plaxton Excalibur	C49FT	1996	
	Volvo B10M-SE	Plaxton Excalibur	C49FT	1996	

Previous Registrations:

1FTG	F907CJW, 245DOC	A3FTG	G720JOG, FTG5	FTG9	From new
2FTG	From new	A4FTG	G727JOG, FTG9	K12FTG	K285XOG
A2FTG	G717JOG, 2FTG	FTG5	From new	L1NER	L11FTG

Livery: Cream, black and silver.

Opposite: **Flight's have been in the news recently having sold their highly successful Flightlink brand to National Express though Flights will continue to provide coaches on a three-year contract. The deal removes Flightlink from the Flight Travel Group's portfolio and passes all ticket handling to National Express. Seen here is L44FTG, a Bova Futura pictured at Hatton Cross and H2FTG, a Toyota Optimo II seen in Wolverhampton.** *Tony Wilson*

GLENSTUART TRAVEL

Glenstuart Travel Ltd, Station Road, Four Ashes,
Wolverhampton WV10 7DB

3	E83OUH	Freight Rover Sherpa	Carlyle Citybus 2	B20F	1987	Ex Victoria Travel, Earlestown, 1995
9	E79OUH	Freight Rover Sherpa	Carlyle Citybus 2	B20F	1987	Ex Victoria Travel, Earlestown, 1995
12	EGB93T	Leyland National 11351A/1R		B52F	1979	Ex Nip-On, St Helens, 1995
15	HNL157N	Leyland National 11351/1R		B52F	1975	Ex Priory Coaches, Gosport, 1994
16	AEF765A	Leyland Leopard PSU3C/4R	Plaxton Supreme III	C49F	1976	Ex Ash, Woburn Moor, 1993
18	RFM887M	Leyland National 1151/1R		B51F	1974	Ex Victoria Travel, Earlestown, 1995
19	MMB973P	Leyland National 11351/1R/SC		DP48F	1976	Ex Evag Cannon, Bolton, 1994
20	UPE204M	Leyland National 1051/1R		B41F	1974	Ex Heaton's Travel, Leigh, 1995

Previous Registrations:
AEF765A   JOX460P

HNL157N is one of six Leyland Nationals operating in the Glenstuart Travel fleet. This model 11351/1R is seen heading for the Pendeford area of Wolverhampton. The livery here is being superseded by a similar base but with the blue line replaced by yellow, red and blue flashes as illustrated in the index section by 16, AEF765A . *Richard Godfrey*

GREEN BUS SERVICE

Warstone Motors Ltd, The Garage, Jacobs Hall Lane, Great Wyrley,
Staffordshire   WS6 6AD

1	NCW151T	Leyland Leopard PSU3E/4R	Duple Dominant	B55F	1979	Ex Vale of Manchester, 1994
2	DDW65V	Leyland Leopard PSU4E/2R	East Lancashire	B45F	1979	Ex Stevensons, 1993
3	TMB877R	Leyland Leopard PSU4D/2R	Duple Dominant	B47F	1976	Ex Parfitt's, Rhymney Bridge, 1994
4	WTJ905L	Leyland Leopard PSU4B/2R	East Lancashire	B45F	1973	Ex Rossendale, 1989
5	URN153V	Leyland Leopard PSU3E/2R	Duple Dominant	B55F	1979	Ex Vale of Manchester, 1994
6	NBZ1676	Leyland Leopard PSU4D/4R	East Lanacshire (1995)	B47F	1976	Ex Rider York, 1992
8	XWG628T	Leyland Atlantean AN68A/1R	Roe	H45/31F	1978	Ex Trans Manche Link, Folkestone, 1993
9	UET678S	Leyland Atlantean AN68A/1R	Alexander AL	H45/31F	1978	Ex Aintree Coachlines, 1990
10	SUA121R	Leyland Atlantean AN68/1R	Roe	H43/33F	1977	Ex Yorkshire Rider, 1995
12	GNY432C	Leyland Titan PD3/4	Massey	L33/35RD	1965	Ex Rhymney Valley, 1981
13	TMB878R	Leyland Leopard PSU4D/2R	Duple Dominant	B47F	1976	Ex Parfitt's, Rhymney Bridge, 1994
14	JUH231W	Leyland Leopard PSU4F/2R	Duple Dominant	B47F	1981	Ex Parfitt's, Rhymney Bridge, 1994
15	GCA747	Bedford OB	Duple Vista	C29F	1950	Ex Sargeant, Llanfaredd, 1973
16	NCW152T	Leyland Leopard PSU3E/4R	Duple Dominant	B55F	1979	Ex Vale of Manchester, 1994
17	NTX576R	Leyland Leopard PSU4C/2R	Willowbrook	B45F	1976	Ex Phil Anslow, Garndiffaith, 1992
18	TMB879R	Leyland Leopard PSU4D/2R	Duple Dominant	B47F	1976	Ex Parfitt's, Rhymney Bridge, 1994
19	DUH77V	Leyland Leopard PSU3E/2R	East Lancashire	B47F	1980	Ex Stevensons, 1993
21	DUH76V	Leyland Leopard PSU3E/2R	East Lancashire	B47F	1980	Ex Stevensons, 1993
22	DUH78V	Leyland Leopard PSU3E/2R	East Lancashire	B47F	1980	Ex Stevensons, 1993
23	YBO17T	Leyland Leopard PSU3E/2R	East Lancashire	B51F	1979	Ex Parfitts, Rhymney Bridge, 1994
24	LUG82P	Leyland Atlantean AN68/1R	Roe	H43/33F	1975	Ex Yorkshire Rider, 1994
27	F249DKG	Freight Rover Sherpa	Carlyle Citybus 2	B20F	1989	Ex JC Mini, Widnes, 1994
28	E149RNY	Freight Rover Sherpa	Carlyle Citybus 2	B20F	1988	Ex Norman Hayes, Connah's Quay, 1994
29	D124WCC	Freight Rover Sherpa	Carlyle	B20F	1987	Ex Worthen Motors, 1994
30	E232NFX	Freight Rover Sherpa	Carlyle Citybus 2	B20F	1987	Ex Midland, 1994

Previous Registrations:
NBZ1676   RWT527R

Livery: Green, cream and yellow

Overleaf: **The Green Bus Services fleet is represented by fleet numbers 7 and 8. Plaxton Derwent-bodied Leyland Leopard 7, RWT531R has recently been taken out of service though was pictured in Wolverhampton last summer. Also seen in Wolverhampton is 8, XWG628T a Leyland Atlantean with Roe bodywork now converted to single door arrangement.**
Richard Godfrey/Tony Wilson

Latterly with Vale of Manchester 5, URN153V was new to Lancaster in 1979. One of the editors thought it might be worth noting that 60% of the fleet listed can claim Welsh ancestry while now working in England.
Richard Godfrey

8 ESSINGTON
8
THE VARSITY
PASSES ACCEPTED
GREEN BUS SERVICE
RWT 531R

BARCLAYS BANK
9
LANDYWOOD
VIA CHESLYN HAY
XWG 628T

HANDYBUS

Matthews Motors & Coach Ltd, The Jumbo Yard, Turner Crescent, Loomer Road Ind Est,
Chesteron, Newcastle-u-Lyme, Staffordshire ST5 7JZ

HTL755N	Bedford YRQ	Plaxton Elite III Express	C45F	1975	Ex Beadles, Newtown, 1995
VRF660S	Ford Transit 160	Deansgate	M12	1977	Ex Morris, Loggerheads, 1993
NBF957V	Ford Transit 160	Dormobile	B16F	1979	Ex Copelands, Stoke-on-Trent, 1995
B413NJF	Ford Transit 190	Rootes	B16F	1985	Ex Irwell Valley, Boothstown, 1992
B414NJF	Ford Transit 190	Rootes	B16F	1985	Ex Irwell Valley, Boothstown, 1992
B422NJF	Ford Transit 190	Rootes	B16F	1985	Ex Stoniers, 1993
C576TUT	Ford Transit 190	Dormobile	B16F	1986	Ex Stevensons, 1992
C89AUB	Ford Transit 160	Carlyle	B18F	1986	Ex Executive Travel, Fenton, 1995
C29WBF	Ford Transit 190	PMT	B16F	1986	Ex Simpson, Kidsgrove, 1995
C544TJF	Ford Transit 160	Rootes	B16F	1986	Ex Morris, Loggerheads, 1993
D113TFT	Freight Rover Sherpa	Carlyle	B18F	1986	Ex Stoniers, 1993
D122TFT	Freight Rover Sherpa	Carlyle	B18F	1986	Ex Potteries Bus Service, Stoke, 1995
D135TFT	Freight Rover Sherpa	Carlyle	B18F	1986	Ex Osborne, West Bromwich, 1993
D101UJC	Freight Rover Sherpa	Dormobile	B16F	1986	Ex Knowles, Ashton-in-Makerfield, 1994
D579EWS	Freight Rover Sherpa	Dormobile	B16F	1986	Ex A1A Travel, Birkenhead, 1995
D413GBF	Renault-Dodge S56	Reeve Burgess Beaver	B23F	1986	Ex Copelands, Stoke-on-Trent, 1995
D423GBF	Renault-Dodge S56	Reeve Burgess Beaver	B23F	1986	Ex Copelands, Stoke-on-Trent, 1995
D4xxGBF	Renault-Dodge S56	Reeve Burgess Beaver	B23F	1986	Ex Copelands, Stoke-on-Trent, 1995
D69NOF	Freight Rover Sherpa	Carlyle	B18F	1987	Ex Midland, 1995
D122WCC	Freight Rover Sherpa	Carlyle	B18F	1987	Ex Stoniers, 1993
G142LRM	Mercedes-Benz 609D	Reeve Burgess Beaver	B20F	1989	Ex North Western, 1995
H731AUE	Ford Transit 160	Ford	M14	1991	Ex Parry, Crewe, 1995

Livery: Blue and red

Previous Registrations:
D423GBF   D157LTA,MIB746   D4..GBF   D152LTA, MIB992   D4..GBF   D167LTA, MIB293
NBF957V   JMJ114V, 555WEH, VXA133, NBF759V, MIB905

A trio of Renault-Dodge S56 minibuses with early Reeve Burgess Beaver bodies new to Plymouth have been acquired by Handybus from Copelands who applied MIB index marks to them. Now with a Stoke-issued index mark, D413GBF is seen at Newcastle bus station. *Tony Wilson*

HAPPY DAYS

Happy Days (Woodseaves) Ltd, Knightley Gorse Garage, Woodseaves,
Staffordshire ST20 0JR
Staffordian Travel Ltd, Greyfriars Coach Station, Greyfriars Way, Stafford ST16 2SH

110	D	Renault-Dodge S56	Northern Counties	B22F	1987	Ex Midland, 1996
111	D559HNW	Iveco Daily 49.10	Robin Hood City Nippy	B24F	1986	Ex Fosseway, Chippenham, 1994
112	D60TLV	Freight Rover Sherpa	Dormobile	B20F	1986	Ex ??
114	D526HNW	Ford Transit 190	Carlyle	B20F	1986	Ex Stevensons, 1994
115	OSR204R	Bristol VRT/LL3/6LXB	Alexander AL	H49/34D	1977	Ex Nottingham, 1994
117	MOD573P	Bristol VRT/SL3/6LXB	Eastern Coach Works	H43/32F	1976	Ex Yelloway, Rochdale, 1989
119	UNW929R	Bristol VRT/SL3/6LXB	Eastern Coach Works	H43/31F	1977	Ex Happy Al's, Birkenhead, 1993
120	RYG385R	Bristol VRT/SL3/6LXB	Eastern Coach Works	H43/31F	1976	Ex Happy Al's, Birkenhead, 1993
121	RYG389R	Bristol VRT/SL3/6LXB	Eastern Coach Works	H43/31F	1976	Ex Happy Al's, Birkenhead, 1993
122	D423NNA	Renault-Dodge S56	Northern Counties	B22F	1987	Ex Midland, 1995
123	FFR169S	Bristol VRT/SL3/6LXB	Eastern Coach Works	DPH43/31F	1978	Ex Nottingham, 1994
133	EBZ5229	Leyland Leopard PSU3E/4R	Duple Dominant	B47F	1977	Ex Blue Bus, Horwich, 1995
134	FWA499V	Leyland Leopard PSU3E/4R	Duple Dominant II Express	C53F	1980	Ex Rannoch Mail & Bus, 1994
139	GIL3167	Leyland Leopard PSU3E/4R	Plaxton Supreme IV	C53F	1979	Ex Swindells, Chadderton, 1994
141	VFV8V	Leyland Leopard PSU3E/4R	Duple Dominant II Express	C53F	1979	Ex Burnley & Pendle, 1995
149	SUX260R	Ford R1114	Duple Dominant	C53F	1977	Ex Jones, Llansilin, 1989
153	YRE465S	Ford R1114	Duple Dominant II	C53F	1978	Ex Greatrex, Stafford, 1985
158	XWX175S	Leyland Leopard PSU3E/4R	Duple Dominant II	C53F	1978	Ex Cooper, Leeds, 1987
166	L300BVA	Volvo B10M-60	Jonckheere Deauville P599	C51FT	1993	
180	SIB6719	Volvo B10M-61	Van Hool Alizée	C49FT	1988	Ex Bluebird of Weymouth, 1995
181	K1HDC	Scania K113CRB	Van Hool Alizée	C48FT	1993	
182	K11HDC	Scania K113CRB	Van Hool Alizée	C48FT	1993	
183	L4WOL	Volvo B10M-60	Plaxton Excalibur	C38FT	1993	
185	L1HDC	Volvo B10M-60	Jonckheere Deauville P599	C51FT	1993	
186	N11HDC	Scania K113CRB	Irizar Century 12.35	C49FT	1996	
187	N63FWU	EOS E180Z	EOS 90	C53FT	1996	

Livery: White, red and black (Happy Days); grey and red (Staffordian) 111/2/24/49/53-9/62/5/6.

Previous Registrations:
SIB6719	E281HRY	EBZ5229	YFR489R
GIL3167	FAG195T, 165DKH, RKH312T		

Double-deck buses are used by Happy Days on school contract work. Seen here is 120, RYG385R a Bristol VRT, and one of three acquired from Happy Al's. Similar bus number 123 is fitted with high back seating.
Bill Potter

Happy Days operate coach contract and tour services from a garage in the Staffordshire countryside near Woodseaves, the bus services having been sold to Midland, the associated Staffordian business is based in the county town. The coach fleet now numbers some 25 vehicles excluding now-sold WHA325, seen here on a Scania chassis with Van Hool Alizée bodywork and L4WOL a Volvo B10M with Plaxton Excalibur bodywork liveried as the Wolverhampton Wanderers' football team coach.
Colin Lloyd/Bill Potter

HARRY SHAW

H Shaw (DM) Ltd, Mill House, Mill Lane, Binley, Coventry   CV3 2DU

Depots : Leicester Street, Bedworth and Mill Lane, Binley

L5URE	Scania K113CRB	Irizar Century	C49FT	1994
L19UER	Scania K113CRB	Irizar Century	C49FT	1994
KOV2	Aüwaerter Neoplan N122/3	Aüwaerter Skyliner	CH57/22CT	1994
L42VRW	Volvo B10M-62	Plaxton Premiére 320	C53F	1994
L43VRW	Volvo B10M-62	Plaxton Premiére 320	C53F	1994
84COV	Volvo B10M-60	Plaxton Premiére 350	C53F	1994
M571BVL	Mercedes-Benz 711D	Autobus Classique	C19F	1994
M211FMR	Mercedes-Benz 711D	Autobus Classique	C24F	1995
M32LHP	Aüwaerter Neoplan N122/3	Aüwaerter Skyliner	CH57/20CT	1995
3KOV	Scania K113TRB	Van Hool Astrobel	CH51/17CT	1995
M34LHP	Scania K113TRB	Irizar Century 12.37	C51FT	1995
HST11	Volvo B10M-62	Plaxton Premiére 350	C49FT	1995
M36LHP	Volvo B10M-62	Plaxton Premiére 350	C49FT	1995
M37LHP	Volvo B10M-62	Plaxton Premiére 350	C49FT	1995
M38LHP	Volvo B10M-62	Plaxton Premiére 350	C49FT	1995
1KOV	Volvo B10M-62	Plaxton Premiére 350	C39FT	1995
M213NHP	Scania K113CRB	Irizar Century 12.35	C49FT	1995
N91WVC	Scania K113TRB	Van Hool Astrobel	CH51/17CT	1996
N92WVC	Scania K113CRB	Van Hool Alizée	C49FT	1996
N93WVC	Scania K113TRB	Irizar Century 12.37	C51FT	1996
N94WVC	Scania K113TRA	Irizar Century 12.37	C51FT	1996
N95WVC	Scania K113CRB	Irizar Century 12.35	C49FT	1996
N96WVC	Volvo B10M-62	Plaxton Premiére 350	C49FT	1996
N97WVC	Volvo B10M-62	Plaxton Premiére 350	C49FT	1996
N98WVC	Volvo B10M-62	Plaxton Premiére 350	C49FT	1996

Previous Registrations:

1KOV	M39LHP	84KOV	L45VRW	KOV2	L41VRW
3KOV	M31LHP	HST11	M35LHP		

Livery: Orange and blue

HI RIDE

F Pellington, 60 Holly Road, Handsworth, Birmingham   B20 2DB

Depot : Park Rose Estate, Middlemore Road, Smethwick

JHA207L	Leyland Leopard PSU3B/2R	Marshall	DP49F	1973	Ex Enterprise & Silver Dawn, 1992
OJD60R	Bristol LH6L	Eastern Coach Works	B39F	1976	Ex Perry, Bromyard, 1992
GWY691N	Leyland Leopard PSU4B/2R	Plaxton Derwent	B43F	1975	Ex Glyn Williams, Crosskeys, 1993

Livery:

The Harry Shaw fleet comprises coaches all new to the company with deliveries currently being divided between Scania and Volvo. One of each are seen here in the form of L43VRW, a Volvo B10M with Plaxton Premiére 350 bodywork and M34LHP a tri-axle Scania with Irizar Century bodywork.
Colin Lloyd

Jones Travelways provide touring services from Market Drayton as well as contract services in the area. Representing the fleet are Dennis Javelin F907UPR with Plaxton Paramount 3200 bodywork seen on tour at Llandudno and J436HDS, a Volvo B10M with the Plaxton Premiére 350 design pictured when visiting the capital. *Ralph Stevens/Colin Lloyd*

HORROCKS

A P Horrocks, Ivy House, Brockton, Lydbury North, Shropshire   SY7 8BA

GAX2C	Bristol RELL6G	Eastern Coach Works	B54F	1965	Ex Wood, Craven Arms, 1992	
WLO574G	Leyland Leopard PSU3A/4R	Plaxton Elite	C51F	1969	Ex Worthen Travel, 1994	
VOD545K	Bristol VRT/SL2/6LX	Eastern Coach Works	H39/31F	1971	Ex Wood, Craven Arms, 1992	
CDC168K	Seddon Pennine VI	Plaxton Elite III	C45F	1972	Ex Trefaldwyn, Montgomery, 1989	
PGX235L	Ford R192	Willowbrook 001	DP43F	1973	Ex Cunningham, Stanford-le-Hope, 1990	
GWP633N	Bedford YRT	Duple Dominant	C53F	1975	Ex Trefaldwyn, Montgomery, 1989	
TTT236R	Bedford YMT	Duple Dominant II	C53F	1977	Ex Trefaldwyn, Montgomery, 1989	
DJF633T	Bedford YMT	Plaxton Supreme IV	C53F	1979	Ex Jones, Annscroft, 1991	
D625BCK	Iveco Daily 49.10	Robin Hood City Nippy	B25F	1987	Ex Ribble, 1993	

Livery: White

JONES

Jones Coachways Ltd, 20a Shropshire Street, Market Drayton, Shropshire   TF9 3BY

Depot: Towers Lawn, Market Drayton

150	XWJ791T	Ford R1114	Plaxton Supreme III	C53F	1979	Ex Finsbury Coaches, 1979
152	JFD296V	Ford R1114	Duple Dominant II	C53F	1979	Ex Olsen, Strood, 1981
155	RAW735X	Ford R1114	Duple Dominant II Express	C53F	1982	
158	A531CUX	Mercedes-Benz L608D	Reeve Burgess	C21F	1984	
159	B250HUX	Ford R1115	Plaxton Paramount 3200 II	C53F	1985	
161w	JNK984N	Ford R1114	Plaxton Elite III	C53F	1975	Ex Goode, West Bromwich, 1985
162	BLJ720Y	Ford R1115	Plaxton Paramount 3200	C53F	1983	Ex Excelsior, 1986
164	D720TNT	Volvo B10M-61	Plaxton Paramount 3500 III	C53F	1987	
165	WNR606S	Ford R1114	Duple Dominant II	C53F	1978	Ex Greenway Travel, Hitchin, 1990
166	G587LUX	Mercedes-Benz 811D	Reeve Burgess Beaver	C25F	1990	
167	G529MNT	Dennis Javelin 12SDA1907	Duple 320	C57F	1990	
168	F907UPR	Dennis Javelin 12SDA1907	Plaxton Paramount 3200 III	C51FT	1989	Ex Goldline, Wimbourne, 1991
169	G92RGG	Volvo B10M-60	Plaxton Paramount 3500 III	C53F	1990	Ex Park's, 1993
170	NLC871V	Volvo B58-61	Plaxton Supreme IV	C51F	1980	Ex Seaview, Parkstone, 1992
171	EDF274T	Leyland Leopard PSU5C/4R	Plaxton Supreme IV	C57F	1979	Ex Spring, Evesham, 1993
172	G93RGG	Volvo B10M-60	Plaxton Paramount 3500 III	C53F	1990	Ex Park's, 1993
173	J436HDS	Volvo B10M-60	Plaxton Premiére 350	C53F	1992	Ex Park's, 1993

Livery: Blue/grey and brown.

KING OFFA

King Offa Travel Services Ltd, The Lodge, Winsley Hall, Westbury,
Shropshire   SY5 9HB

DNT527T	Leyland Leopard PSU3E/4R	Plaxton Supreme III Express	C50F	1978	Ex Midland Red North, 1990
WOC739T	Leyland Leopard PSU3E/4R	Plaxton Supreme III Express	C49F	1979	Ex Midland Red North, 1992
DDM22X	Leyland Leopard PSU3F/4R	Willowbrook 003	C53F	1981	Ex Grimsby Cleethorpes, 1994
MUV837X	Leyland Leopard PSU5C/4R	Duple Dominant IV	C53F	1982	Ex Midland Red South, 1994
HKP126	Leyland Tiger TRCTL11/3R	Plaxton Paramount 3500	C49FT	1983	Ex Drawlane, 1991
A101JJT	Leyland Tiger TRCTL11/3R	Plaxton Paramount 3200	C57F	1984	Ex Midland Red North, 1992
M551ONT	Dennis Dart 9.8SDL3040	Plaxton Pointer	B40F	1994	

Previous Registrations:
DNT527T   APR818T, HKP126   HKP126   BRN4Y

Livery: White, blue and red.

KNOTTY

MA & SM Hearson, Unit C, Parkhouse Road East, Parkhouse Ind Est, Chesterton,
Newcastle-under-Lyme, Staffordshire   ST5 7RB

9	UAD316H	Daimler Roadliner SRP8	Plaxton Elite	C47F	1970	Ex Offerton Coaches, Stockport, 1989
19	CYA181J	AEC Reliance 6MU3R	Plaxton Derwent	B47F	1970	Ex Chiltern Queens, Woodcote, 1992
	DVT167J	AEC Reliance 6U2R	Alexander Y	DP49F	1970	Ex Anderson, Iver, 1995
24	EDJ242J	AEC Swift 2MP2R	Marshall	B44D	1971	Ex preservation, 1994
17	JPA171K	AEC Reliance 6U2R	Park Royal	DP45F	1972	Ex Buffalo, Flitwick, 1992
23	JPF103K	AEC Swift 3MP2R	Alexander W	DP48F	1972	Ex Blue Triangle, Bootle, 1993
	CRP310K	Leyland Leopard PSU3B/4R	Plaxton Elite III	C53F	1972	Ex Evans, Tregaron, 1995
	JWO891L	AEC Reliance 6MU4R	Plaxton Elite III Express	C51F	1973	Ex Hall, Biddulph, 1996
	TIB2893	AEC Reliance EBC	Berkhof Esprit 340(1984)	C49F	1973	Ex Crabtree, Bradford, 1994
20	OFR983M	AEC Swift 3MP2R	Marshall	B47D	1973	Ex Wealden Beeline, 1993
25	NKN101M	AEC Reliance 6U3ZR	Plaxton Elite III	C53F	1975	Ex Davies, Hawkhurst, 1994
3	KVE909P	AEC Reliance 6U3ZR	Plaxton Supreme III Express	C49F	1975	Ex Premier, Cambridge, 1988
	LHS480P	AEC Reliance 6MU4R	Willowbrook	B45F	1975	Ex Owen, Cappenhall, 1991
21	MWA839P	AEC Reliance 6U3ZR	Plaxton Supreme III	C53F	1976	Ex Fallon, Dunbar, 1993
2	PPH431R	AEC Reliance 6U3ZR	Plaxton Supreme III Express	C53F	1977	Ex LMS Travel, Chesterton, 1991
	XWX176S	Leyland Leopard PSU3E/4R	Duple Dominant II	C53F	1977	Ex Evans, Tregaron, 1996
10	TPD12S	AEC Reliance 6U2R	Plaxton Supreme III Express	C53F	1977	Ex Happy Times, Wednesfield, 1990
	APM106T	AEC Reliance 6U2R	Plaxton Supreme IV Express	C53F	1979	Ex Gary's, Tredegar, 1996
14	ANA8T	AEC Reliance 6U3ZR	Plaxton Supreme III	C53F	1978	Ex Lancaster, 1991
	WEB407T	AEC Reliance 6U3ZR	Plaxton Supreme III Express	C49F	1978	Ex Stagecoach Cambus, 1996
	WEB408T	AEC Reliance 6U3ZR	Plaxton Supreme IV Express	C49F	1978	Ex Stagecoach Cambus, 1996
	WEB410T	AEC Reliance 6U3ZR	Plaxton Supreme IV Express	C49F	1978	Ex Stagecoach Cambus, 1996
	WEB411T	AEC Reliance 6U3ZR	Plaxton Supreme IV Express	C49F	1978	Ex Stagecoach Cambus, 1996
	FBC473T	AEC Reliance 6U3ZR	Plaxton Supreme IV	C57F	1979	Ex Young, Blackwell, 1995
	IIL6436	Bedford YMT	Unicar	C53F	1980	Ex Brough, Norton, 1995

Livery: White, grey, black and red.

Previous Registrations:
TIB2893   HFN53L, A198TAR   IIL6436   LBC935V, FIL6657, MBC218V
FBC473T   EBM455T, 481DRC

Opposite: **Two local operators in the area have become known for their attractive liveries.** *Above* **is the Plaxton Pointer-bodied Dennis Dart M551ONT used by King Offa on Shropshire Bus services.** *Below* **is JPA171K of Knotty, the operator of predominantly AEC vehicles of various models and bodywork. This Park-Royal-bodied dual-purpose vehicle was once part of a large batch delivered for Green Line service to the then newly formed London Country.**   *Richard Godfrey/Cliff Beeton*

822 TELFORD
KING OFFA
KING OFFA
M551 ONT
DENNIS
PLAXTON

BRADELEY
KNOTTY
Reliance
JPA 171K
KNOTTY

LEON'S OF STAFFORD

Leon's Coach Travel (Stafford) Ltd, Redhill Garage, First Avenue,
Stafford   ST16 0RY

Depots : Ludford Fields Ind Est, Seighford and Fisrt Avenue, Stafford

27	VRF566X	Mercedes-Benz L207D	Whittaker	M12	1982	
34	B192PFA	Mercedes-Benz L608D	PMT Hanbridge	C21F	1985	
35	TEH377W	Volvo B10M-61	Plaxton Viewmaster IV	C57F	1981	Ex Price, Halesowen, 1985
38	LOI8643	Volvo B10M-61	Plaxton Paramount 3500	C53F	1983	Ex Dave Parry, Cheslyn Hay, 1986
39	LOI9772	Volvo B10M-61	Van Hool Alizée	C49FT	1984	Ex Park's, 1986
43	A21FVT	Mercedes-Benz L508D	Reeve Burgess	C18F	1983	Ex Midland Red North, 1987
48	E590LEH	Mercedes-Benz 811D	Optare StarRider	C29F	1988	
50	LOI7191	Volvo B10M-61	Van Hool Alizée	C53FT	1989	
54	4327PL	Scania K113CRB	Plaxton Paramount 3500 III	C49FT	1989	
55	LOI1454	Volvo B10M-60	Ikarus Blue Danube	C49FT	1990	
56	CRE240T	Ford Transit 160	Deansgate	M12	1978	
59	8636PL	Scania K93CRB	Duple 320	C55F	1990	
60	H330JVT	Volvo B10M-60	Plaxton Paramount 3500 III	C49FT	1990	
61	NDH7P	Ford R1014	Duple Dominant	C45F	1976	Ex Taylor, Derby, 1990
62	C230AEA	Ford Transit VE	Ford	M8	1986	
65	LJI8160	Volvo B10M-61	Plaxton Paramount 3500	C48FT	1988	Ex Flight's, 1991
68	J65SRE	Ford Transit VE6	Deansgate	M14	1992	

Leon's tri-axle Scania number 83, M1LCT is seen in this attractive livery in Trafalgar Square while escorting a party of Australian tourists.   The Van Hool Alizée-bodied coach features full air-conditioning and many other extras. *Colin Lloyd*

The more local operations of Leon's are represented by 85, BPT922S, a Bristol VR acquired from Northumbria. This joined a former WMPTE VR and two Fleetlines formerly with Nottingham for school contracts, although one of the latter has since been withdrawn. *Tony Wilson*

70	J62SRE	Volvo B10M-60	Plaxton Paramount 3500 III	C50FT	1992	
71	E752XHL	Mercedes-Benz 609D	Whittaker Europa	C24F	1988	Ex Wickson, Clayhanger, 1992
73	K680BRE	Scania K113CRB	Plaxton Paramount 3500 III	C49FT	1993	
74	5888EH	Volvo B10M-61	Plaxton Paramount 3500	C49FT	1983	Ex Bushell, Burton, 1993
75	L2LCT	Volvo B10M-60	Jonckheere Deauville P599	C51FT	1993	
76	G117OGA	Mercedes-Benz 811D	Optare StarRider	C29F	1990	Ex Broomfield, Hawick, 1994
77	5702PL	Scania K92CRB	Van Hool Alizée	C55F	1989	Ex PMT, 1994
78	OIA1652	Scania K93CRB	Van Hool Alizée	C55FT	1990	Ex PMT, 1994
80	2335PL	Scania K112CRB	Van Hool Alizée	C51FT	1988	Ex PMT, 1994
81	9346PL	Scania K93CRB	Van Hool Alizée	C51FT	1990	Ex PMT, 1994
83	M1LCT	Scania K113TRA	Van Hool Alizée	C49FT	1994	
84	M2LCT	Scania K113CRB	Irizar Century 12.35	C49FT	1995	
85	BPT922S	Bristol VRT/SL3/6LXB	Eastern Coach Works	H43/31F	1977	Ex Northumbria, 1995
87	GOG684N	Bristol VRT/SL2/6LX	MCW	H43/33F	1975	Ex Porthcawl Omnibus, 1994
88	UTV224S	Leyland Fleetline FE30AGR	Northern Counties	H47/31D	1978	Ex Nottingham, 1995
89	HIL6245	Volvo B10M-61	Plaxton Paramount 3500 III	C49FT	1988	Ex Cedric's, Wivenhoe, 1995
90	PNH184	Volvo B10M	Plaxton Paramount 3500 III	C49FT	1988	Ex Cedric's, Wivenhoe, 1995
91	A712RCA	Mercedes-Benz L207D	Imperial	M12	1983	Ex Lockley Cs, Stafford, 1996
92	C588UFA	Ford Transit VE6	Mellor	C16F	1985	Ex Alpine Taxis, Weston-s-Mare, 1996
93	K892BSX	Ford Transit VE6	Ford	M14	1992	Ex Munro, Uddingston, 1996
94	N2LCT	Scania K113CRB	Irizar Century	C49FT	1996	
95	N3LCT	Mercedes-Benz 814D	Auto Classique Nouvelle	C29F	1996	

Previous Registrations:

2335PL	E516YWF	LOI1454	G448BVT
4327PL	G360XEH	LOI7191	F150SRF
5702PL	F115UEH	LOI8643	A111GUE
5888EH	FDH294Y, 1398NT, XUJ435Y	LOI9772	A644UGD
8636PL	G655EFA	OIA1652	G806FJX
9346PL	G805FJX	PNH184	E747JAY
HIL6245	E906UNW	TEH377W	DUY596W, LOI1454
LJI8160	E907UNW		

Livery: Cream and red.

The North & West Midlands Bus Handbook 47

Pete's Travel, Lionspeed and Busy Bus share a fleet liveried in green and yellow and are most commonly seen in West Bromwich. That town's bus station is the location of the pictures. Above is F889XOE, a Freight Rover Sherpa with Carlyle Citybus 2 bodywork while below is D409NUH, a Dodge G08 with high-back seated East Lancashire bodywork new to Islwyn. *Richard Godfrey*

LIONSPEED / PETE'S

Petes Travel Ltd; Lionspeed Ltd; Busy Bus Co, Unit 52 Queens Court Trad Est,
Greats Green, West Bromwich, West Midlands B70 9EL

Depots : Coventry Road, Birmingham; Brookside, Great Barr; Metalloys Ind Est, Minworth and Queens Court Trading Estate, West Bromwich.

C817CBU	Renault-Dodge S56	Northern Counties	B18F	1986	Ex Manchester Airport, 1995
C824CBU	Renault-Dodge S56	Northern Counties	B18F	1986	Ex Manchester Airport, 1995
C832CBU	Renault-Dodge S56	Northern Counties	B18F	1986	Ex Manchester Airport, 1995
C833CBU	Renault-Dodge S56	Northern Counties	B18F	1986	Ex McAinsh, Reading, 1993
C834CBU	Renault-Dodge S56	Northern Counties	B18F	1986	Ex Manchester Airport, 1995
D669SEM	Renault-Dodge S56	Northern Counties	B22F	1986	Ex PMT, 1995
D858LND	Renault-Dodge S56	Northern Counties	B18F	1986	Ex PMT, 1995
D138LTA	Renault-Dodge S56	Reeve Burgess	B23F	1986	Ex Cardiff Bus, 1993
D158LTA	Renault-Dodge S56	Reeve Burgess	B23F	1986	Ex Cardiff Bus, 1994
D161LTA	Renault-Dodge S56	Reeve Burgess	B23F	1986	Ex Cardiff Bus, 1994
D175LTA	Renault-Dodge S56	Reeve Burgess	B23F	1986	Ex Martins, Cold Meece, 1993
D257OOJ	Freight Rover Sherpa	Carlyle	B18F	1987	Ex PMT, 1995
D259OOJ	Freight Rover Sherpa	Carlyle	B18F	1987	Ex PMT, 1995
D822PUK	Freight Rover Sherpa	Carlyle	B18F	1987	Ex Bentley, Birmingham, 1993
D828PUK	Freight Rover Sherpa	Carlyle	B18F	1987	Ex Bentley, Birmingham, 1993
D414FEH	Freight Rover Sherpa	PMT	B20F	1987	Ex PMT, 1995
D416FEH	Freight Rover Sherpa	PMT	B20F	1987	Ex PMT, 1995
D126OWG	Renault-Dodge S56	Reeve Burgess	B25F	1987	Ex Stoniers, Newcastle, 1993
D142RAK	Renault-Dodge S56	Reeve Burgess	B25F	1987	Ex Moffatt & Williamson, Gauldry, 1994
D148RAK	Renault-Dodge S56	Reeve Burgess	B25F	1987	Ex Moffatt & Williamson, Gauldry, 1994
D169RAK	Renault-Dodge S56	Reeve Burgess	B25F	1987	Ex Clydeside, 1995
D821RYS	Renault-Dodge S56	Alexander AM	B25F	1987	Ex Martins, Cold Meece, 1993
D39NDV	Renault-Dodge S56	East Lancashire	B24F	1987	Ex Islwyn, 1995
D974TKC	Renault-Dodge S56	Northern Counties	B22F	1987	Ex PMT, 1995
D935NDB	Renault-Dodge S56	Northern Counties	B20F	1987	Ex Little Red Bus, 1995
D710TWM	Renault-Dodge S56	Northern Counties	B22F	1987	Ex Clydeside, 1995
D711TWM	Renault-Dodge S56	Northern Counties	B25F	1987	Ex City Buslines, Birmingham, 1995
D713TWM	Renault-Dodge S56	Northern Counties	B25F	1987	Ex Clydeside, 1995
D870MDB	Renault-Dodge S56	Northern Counties	B20F	1987	Ex Little Red Bus, Smethwick, 1995
D943NDB	Renault-Dodge S56	Northern Counties	B20F	1987	Ex Little Red Bus, Smethwick, 1995
D409NUH	Dodge G08	East Lancashire	DP25F	1987	Ex Islwyn, 1995
D410NUH	Dodge G08	East Lancashire	DP25F	1987	Ex Islwyn, 1995
E727HBF	Freight Rover Sherpa	PMT	B20F	1987	Ex PMT, 1995
E728HBF	Freight Rover Sherpa	PMT	B20F	1987	Ex PMT, 1995
E87OUH	Freight Rover Sherpa	Carlyle Citybus 2	B20F	1987	Ex Shamrock, Pontypridd, 1994
E176UWF	Renault-Dodge S56	Reeve Burgess	B25F	1987	Ex Clydeside, 1995
E185UWF	Renault-Dodge S56	Reeve Burgess	B25F	1987	Ex Clydeside, 1995
E783SJA	Renault-Dodge S56	Northern Counties	B20F	1987	Ex Little Red Bus, Smethwick, 1995
E785SJA	Renault-Dodge S56	Northern Counties	B20F	1987	Ex Little Red Bus, Smethwick, 1995
E269BRG	Renault-Dodge S56	Alexander AM	DP19F	1987	Ex Red & White, 1994
E316NSX	Renault-Dodge S56	Alexander AM	B25F	1988	Ex Victoria Travel, Earlestown, 1994
F889XOE	Freight Rover Sherpa	Carlyle Citybus 2	B20F	1988	Ex The Wright Company, Wrexham, 1994
F895XOE	Freight Rover Sherpa	Carlyle Citybus 2	B20F	1988	Ex The Wright Company, Wrexham, 1994
F233BAX	Freight Rover Sherpa	Carlyle Citybus 2	B18F	1988	Ex Cynon Valley, 1993
N627BWG	Mercedes-Benz 811D	Mellor	B31F	1995	
N628BWG	Mercedes-Benz 811D	Mellor	B31F	1995	
N629BWG	Mercedes-Benz 811D	Mellor	B31F	1995	

Livery: Yellow and green

LITTLE RED BUS

Little Red Bus Co Ltd; Red Arrow Express Ltd, 89 Rabone Lane,
Smethwick, West Midlands B66 3JJ

Depot : Park Rose Estate, Middlemore Road, Smethwick.

EX9779	AEC Reliance MU3RV	Duple Britannia	C41F	1956	Ex Staniforth, Birmingham, 1993
XHA875	Volvo B58-61	Plaxton Viewmaster IV	C53F	1981	Ex M&J Travel, Shrewsbury, 1995
JTC993X	Renault-Dodge S66	Reeve Burgess	B21F	1982	Ex Clun Valley, Newcastle, 1995
C819CBU	Renault-Dodge S56	Northern Counties	B18F	1986	Ex G M Buses, 1992
C822CBU	Renault-Dodge S56	Northern Counties	B18F	1986	Ex G M Buses, 1992
D107OWG	Renault-Dodge S56	Reeve Burgess Beaver	B25F	1987	Ex South Yorkshire, 1992
D707TWM	Renault-Dodge S56	Northern Counties	B22F	1987	Ex Merseybus, 1993
D712TWM	Renault-Dodge S56	Northern Counties	B22F	1987	Ex Merseybus, 1993
D715TWM	Renault-Dodge S56	Northern Counties	B22F	1987	Ex Merseybus, 1993
D246VNL	Renault-Dodge S56	Alexander AM	DP19F	1986	Ex Red & White, 1994
D255VNL	Renault-Dodge S56	Alexander AM	DP19F	1987	Ex Red & White, 1994
D257YBB	Renault-Dodge S56	Alexander AM	DP19F	1987	Ex Red & White, 1994
D871MDB	Renault-Dodge S56	Northern Counties	B18F	1987	Ex G M N, 1995
D914NDB	Renault-Dodge S56	Northern Counties	B18F	1987	Ex G M N, 1995
D918NDB	Renault-Dodge S56	Northern Counties	B18F	1987	Ex G M N, 1995
D919NDB	Renault-Dodge S56	Northern Counties	B18F	1987	Ex G M N, 1995
D925NDB	Renault-Dodge S56	Northern Counties	B18F	1987	Ex G M N, 1995
D928NDB	Renault-Dodge S56	Northern Counties	B18F	1987	Ex G M N, 1995
D929NDB	Renault-Dodge S56	Northern Counties	B18F	1987	Ex G M N, 1995
D935NDB	Renault-Dodge S56	Northern Counties	B18F	1987	Ex G M N, 1995
D936NDB	Renault-Dodge S56	Northern Counties	B18F	1987	Ex G M N, 1995
D937NDB	Renault-Dodge S56	Northern Counties	B18F	1987	Ex G M N, 1995
D946NDB	Renault-Dodge S56	Northern Counties	B18F	1987	Ex G M N, 1995
D972PJA	Renault-Dodge S56	Northern Counties	B18F	1987	Ex G M N, 1995
D976PJA	Renault-Dodge S56	Northern Counties	B18F	1987	Ex G M N, 1995
D780RBU	Renault-Dodge S56	Northern Counties	B18F	1988	Ex G M N, 1995
E277BRG	Renault-Dodge S56	Alexander AM	B19F	1987	Ex Red & White, 1994
E278BRG	Renault-Dodge S56	Alexander AM	DP19F	1987	Ex Red & White, 1994
YHA116	Bedford YMP	Plaxton Paramount 3200 III	C35F	1987	Ex McLaughlin, Penwortham, 1994
E635DCK	Renault-Dodge S46	Dormobile	B25F	1987	Ex United Counties, 1993
E638DCK	Renault-Dodge S46	Dormobile	B25F	1987	Ex United Counties, 1993
E639DCK	Renault-Dodge S46	Dormobile	B25F	1987	Ex United Counties, 1993
E645DCK	Renault-Dodge S46	Dormobile	B25F	1987	Ex United Counties, 1993
E796SJA	Renault-Dodge S56	Northern Counties	B18F	1987	Ex G M N, 1995
E989SJA	Renault-Dodge S56	Northern Counties	B18F	1988	Ex G M N, 1995
E991SJA	Renault-Dodge S56	Northern Counties	B18F	1988	Ex G M N, 1995

Previous Registrations:

EX9779	From new	YHA116	D793SGB, FTG5, D949CFR
XHA875	KHL559W, LTH742		

Livery: Red

Now with the Little Red Bus operation are four Renault-Dodge S56 with Dormobile bodywork latterly operating for United Counties though new to Ribble. E638DCK is seen in Carrs Lane, Birmingham where it is seen heading for the city centre.
Tony Wilson

LONGMYND

T G Evans, Britannia Bank Garage, Pontesbury, Shropshire SY5 0QG

JRB914N	Bedford YRQ	Plaxton Elite III	C45F	1975	Ex O'Brien, Farnworth, 1980
BTU565S	Bedford YLQ	Plaxton Supreme III	C45F	1977	Ex Hanmers, Wrexham, 1980
NVR907W	Volvo B58-61	Plaxton Supreme IV	C57F	1980	Ex TRJ, Golbourne, 1984
LRC21W	Volvo B58-56	Plaxton Supreme IV	C53F	1980	Ex Luxicoaches, Borrowash, 1985
UAB943Y	Volvo B10M-61	Jonckheere Jubilee P90	C53FT	1983	Ex DRM, Bromyard, 1990
TGE93	Volvo B10M-61	Jonckheere Jubilee P90	C51FT	1984	Ex West Kingsdown Coaches, 1986
C377MAW	Volvo B10M-61	Plaxton Paramount 3500 II	C49FT	1986	
E504KNV	Volvo B10M-61	Jonckheere Jubilee P599	C57F	1988	Ex Tellings-Golden Miller, 1991
E170OMD	Volvo B10M-61	Plaxton Paramount 3200 III	C57F	1988	Ex Frames Rickards, Brentford, 1992
E311OMG	Volvo B10M-61	Plaxton Paramount 3200 III	C53F	1988	Ex London Cityrama, 1994
F584BAW	Volvo B10M-61	Jonckheere Jubilee P599	C51FT	1988	
F969HGE	Volvo B10M-61	Plaxton Paramount 3200 III	C57F	1989	Ex Park's, 1990
F210PNR	Toyota Coaster HB31R	Caetano Optimo	C21F	1989	Ex Horseshoe, Tottenham, 1990
G30POD	Volkswagen Transporter	Devon Conversions	M10	1989	
G839GNV	Volvo B10M-60	Jonckheere Deauville P599	C51FT	1989	Ex Inland Travel, Flimwell, 1992
G160XJF	MAN 10-180	Caetano Algarve	C35F	1990	Ex Clegg & Brooking, Middle Wallop, 1995
K1TGE	Volvo B10M-60	Jonckheere Deauville P599	C51FT	1993	

Previous Registrations:
TGE93   A118SNH   UAB943Y   NVV552Y, MOI3565

Livery: White, red and black.

The Belgian coachbuilder, Jonckheere, built six of the Longmynd Volvo's. Early imports were supplied via Roeslare Sales in Northampton, later returning to that town under their own name. Now the products are available through the Volvo-owned Yeates dealership. E504KNV with a P599 Jubilee body dates from the middle period.
Bill Potter

LUDLOWS

Ludlows of Halesowen Ltd, 239 Stourbridge Road, Halesowen,
West Midlands B63 3QU

Depot : Coombes Road, Halesowen

ABW210L	Leyland National 1151/1R/0401		B52F	1973	Ex Tappins, Didcot, 1992
NPD111L	Leyland National 1151/2R/0402		B46F	1972	Ex City Line, 1989
NAT222A	Leyland National 1151/1R/0402		B49F	1973	Ex Eastern National, 1989
NAT333A	Leyland National 1051/1R/0502		B44F	1973	Ex Southern Vectis, 1987
NAT555A	Leyland National 1151/1R/0401		B52F	1973	Ex Eastern National, 1989
PVT244L	Leyland National 1151/1R/0401		B52F	1973	Ex PMT, 1990
WNO556L	Leyland National 1151/1R/0401		B52F	1973	Ex Eastern National, 1989
TPD194M	Leyland National 1051/1R		B41F	1974	Ex Southend, 1992
TPD195M	Leyland National 1051/1R		B41F	1973	Ex Southend, 1992
RFM886M	Leyland National 1151/1R		B49F	1974	Ex Crosville, 1987
ORP466M	Leyland National 1151/1R		B52F	1974	Ex United Counties, 1988
UPE199M	Leyland National 1051/1R		B41F	1974	Ex Southend, 1992
GHU645N	Leyland National 10351/1R		B41F	1975	Ex Southend, 1992
KCR108P	Leyland National 10351/2R		B40D	1976	Ex Orion, Kircaldy, 1993
OFD231P	Leyland National 11351/1R		B50F	1976	Ex Athelstan, Malmesbury, 1989
PTT80R	Leyland National 11351A/1R		B52F	1976	Ex Athelstan, Malmesbury, 1989
VPT596R	Leyland National 11351A/1R		B49F	1977	Ex Tyneside, 1991
NWO460R	Leyland National 11351A/1R/SC		DP48F	1977	Ex Rhondda, 1994
JBR690T	Leyland National 10351A/1R (Volvo)		B49F	1978	Ex Dent, Market Rasen, 1993
BOK62T	Leyland Leopard PSU3E/4R	Plaxton Supreme IV	C53F	1979	Ex Rover, Horsley, 1992
BYW360V	Leyland National 10351A/2R		B44F	1979	Ex Parfitt's, Rhymney Bridge, 1995
DSJ307V	Volvo B58-56	Duple Dominant II	C53F	1980	Ex Arvonia, Llanrug, 1992
YWD687	Leyland TRCTL11/3R	Duple Dominant IV	C51F	1985	Ex Safeguard, Guildford, 1992
SXD696	Volvo B10M-61	Van Hool Alizée	C53F	1986	Ex Shearings, 1991
1223PL	Volvo B10M-61	Ikarus Blue Danube	C53F	1989	Ex Alexanders, Aberdeen, 1989
8797PL	Scania K113CRB	Van Hool Alizée	C49F	1990	Ex Happy Days, Woodseaves, 1994
H720LOL	Dennis Dart 9SDL3002	Carlyle Dartline	B36F	1990	
J272SOC	Dennis Dart 9.8SDL3004	Carlyle Dartline	B43F	1991	

Previous Registrations:

1223PL	F108SSE	NAT333A	XDL802L
8797PL	G554CRF, WHA325, G331FRF	NAT555A	WNO551L
ABW210L	LWN713L, 653GBU	OFD231P	MOD818P, CSU992
BOK62T	CTM413T, YWD687	SXD696	C535DND
KCR108P	KCR108P, 2704MAN		
NAT222A	KCG610L	YWD687	B717MPC

Livery: White; Blue, yellow, white and red (Coaches).

Representing the twenty Leyland Nationals in Ludlows fleet is NAT222A, previously KCG610L and one of the early models new to Alder Valley. *Philip Lamb*

52

M&J TRAVEL

WM & JM Price, Coach Garage, Newcastle, Shrewsbury
Shropshire   SY7 8QL

HLG812K	Bedford VAL70	Plaxton Elite II	C53F	1972	Ex Lakelin, Felindre, 1989
PJT205R	Bedford YMT(Cummins)	Plaxton Supreme III	C53F	1976	Ex Beeline, Warminster, 1996
A512LPP	Bedford YMQ	Plaxton Paramount 3200	C35F	1983	Ex Miller, Box, 1993
A845GEJ	Bedford YNT	Plaxton Paramount 3200	C53F	1984	Ex Evans, Tregaron, 1995
C43WBF	Ford Transit 190	Dormobile	B16F	1986	Ex Midland, 1995
D569RKW	Beford YNV Venturer	Plaxton Paramount 3200 II	C57F	1987	Ex N&R, Elsecar, 1995
D945OKK	Peugeot-Talbot Express	Talbot	M14	1987	Ex Bennett, Bournemouth, 1991
E98LBC	Dennis Javelin 11SDA1906	Duple 320	C53F	1988	Ex Evans, Tregaron, 1993
?	LAG G355Z	LAG Panoramic	C49FT	1989	Ex Supreme, Coventry, 1996
G864MYA	Dennis Javelin 12SDA1907	Duple 320	C53FT	1990	Ex Coombs, Weston-s-Mare, 1996
J416UOC	Leyland DAF 200	Leyland-DAF	M12	1991	Ex Blythewood, Glasgow, 1996
L53JUJ	LDV 400	LDV	M16	1995	
N110VAW	LDV 200	LDV	M11	1995	

Previous Registrations:
A845GEJ   A34UGA, 4012VC   PJT205R   OTX41R, 419CLB, TAP461R, FJW518

Livery: White, black and gold.

Cardiff is the location for this picture of M&J Travel's E98LBC, an early Dennis Javelin with Duple 320 bodywork. *Richard Eversden*

MERRY HILL MINI

Merry Hill Minibuses Ltd, 100 Dudley Road East, Oldbury, West Midlands   B69 3DY

110-138			Freight Rover Sherpa		Carlyle Citybus 2		B20F	1988		
110	E514TOV	129	F879XOE	133	F883XOE	135	F885XOE	137	F887XOE
127	F877XOE	131	F881XOE	134	F884XOE	136	F886XOE	138	F888XOE
128	F878XOE	132	F882XOE						

140	H723LOL	Freight Rover Sherpa	Carlyle Citybus 2	B20F	1990	
141	G227EOA	Freight Rover Sherpa	Carlyle Citybus 2	B20F	1990	Ex Skills, Nottingham, 1991
142	G228EOA	Freight Rover Sherpa	Carlyle Citybus 2	B25F	1989	Ex Strathclyde Buses, 1991
143	E200TVE	MCW MetroRider MF150/82	MCW	B25F	1988	Ex Richardson, Oldbury, 1991
144	H713LOL	Freight Rover Sherpa	Carlyle Citybus 2	B20F	1990	Ex Strathclyde Buses, 1992
145	H714LOL	Freight Rover Sherpa	Carlyle Citybus 2	B20F	1990	Ex Strathclyde Buses, 1991

146-153			Freight Rover Sherpa		Carlyle Citybus 2		B20F	1988-89 Ex National Welsh, 1992-93	
146	G273HBO	148	G264GKG	150	G267GKG	152	G271GKG	153	G272GKG
147	G263GKG	149	G265GKG	151	G270GKG				

154-175			Optare MetroRider		Optare		B23F	1995-96	
154	N468WDA	159	N94WOM	164	N811XOJ	168	N149BOF	172	N154BOF
155	N469WDA	160	N95WOM	165	N812XOJ	169	N151BOF	173	N155BOF
156	N91WOM	161	N268XOJ	166	N147BOF	170	N152BOF	174	N156BOF
157	N92WOM	162	N269XOJ	167	N148BOF	171	N153BOF	175	N157BOF
158	N93WOM	163	N270XOJ						

Livery: Silver and black or white

During 1995 Merry Hill Mini started to take delivery of Optare MetroRiders and delivery has continued steadily throughout the winter. One of the 1996 intake is 159, N94WOM, seen here in Dudley while heading for the shopping centre. *Tony Wilson*

METROPOLITAN

Buslink (Midlands) Ltd, Orchard Cottage, Stubwood, Uttoxeter   ST14 5MX

Depot : Rowney House Ind Est, Darlaston

C725JJO	Ford Transit 190	Carlyle	B18F	1986	Ex Stevenson, 1994
D302SDS	Renault-Dodge S56	Alexander AM	B25F	1986	Ex Western Scottish, 1995
D304SDS	Renault-Dodge S56	Alexander AM	B25F	1986	Ex South Lancs, St Helens, 1995
D307SDS	Renault-Dodge S56	Alexander AM	B25F	1986	Ex South Lancs, St Helens, 1995
D303MHS	Renault-Dodge S56	Alexander AM	B21F	1986	Ex Mainline, 1995
D314MHS	Renault-Dodge S56	Alexander AM	B21F	1986	Ex Mainline, 1995
D849LND	Renault-Dodge S56	Northern Counties	B20F	1986	Ex City Buslines, Birmingham, 1995
D222NCS	Renault-Dodge S56	Alexander AM	B25F	1987	Ex Western Scottish, 1995
D224NCS	Renault-Dodge S56	Alexander AM	B25F	1987	Ex Western Scottish, 1995
D231NCS	Renault-Dodge S56	Alexander AM	B25F	1987	Ex Western Scottish, 1995
D234NCS	Renault-Dodge S56	Alexander AM	B25F	1987	Ex Western Scottish, 1995
D256NCS	Renault-Dodge S56	Alexander AM	B25F	1987	Ex Western Scottish, 1995
D153RAK	Renault-Dodge S56	Reeve Burgess	B25F	1987	Ex Mainline, 1995
D169RAK	Renault-Dodge S56	Reeve Burgess	B25F	1987	Ex Mainline, 1995
D431TCA	Renault-Dodge S56	Northern Counties	B22F	1987	Ex Alpine, Llandudno, 1995
D432TCA	Renault-Dodge S56	Northern Counties	B22F	1987	Ex Alpine, Llandudno, 1995
D433TCA	Renault-Dodge S56	Northern Counties	B22F	1987	Ex Alpine, Llandudno, 1995
D706TWM	Renault-Dodge S56	Northern Counties	B25F	1987	Ex South Lancs, St Helens, 1995
G737NNS	Mercedes-Benz 811D	Wright Nim-Bus	B29F	1990	Ex Timeline, 1995
H552SWE	Mercedes-Benz 811D	Whittaker	B29F	1990	Ex City Traveller, Hull, 1994
J648XHL	Dennis Dart 9.8SDL3004	Reeve Burgess Pointer	B40F	1991	Ex Star Line, Knutsford, 1995
M801OJW	Dennis Dart 9.8SDL3040	Plaxton Pointer	B40F	1995	

Livery: Cream and green

Services between Dudley and the Merry Hill shopping centre are provided by several operators among them Metropolitan, a recent arrival on the service. Departing Dudley bus station is Metropolitan's newest vehicle, M801OJW, a Dennis Dart with Plaxton Pointer bodywork.   *Tony Wilson*

MIDLAND

Midland Red (North) Ltd, Delta Way, Longford Road, Cannock,
Staffordshire WS11 3XB

Depots & outstations: Abermule; Bridgnorth; Delta Way, Cannock; Bus Station, Crewe; Oswald Road, Oswestry; Ditherington, Shrewsbury, Donnington Ind Park, Common Lane, Stafford; Aldergate, Tamworth; Charlton Street, Wellington and Woodseaves, Stafford.

37	C37WBF	Ford Transit 190D	Dormobile	B16F	1986			
88-110		Ford Transit VE6	Dormobile	B16F	1986			
88	D88CFA	95	D95CFA	98	D98CFA	104	D104CFA	107	D107CFA
91	D91CFA	96	D96CFA	102	D102CFA	105	D105CFA	110	D110CFA
92	D92CFA								
180	F39HOD	Ford Transit VE6	Dormobile	B18F	1988	Ex Panda Hire, Exeter, 1990		
181-191		Ford Transit VE6	Dormobile	B18F	1990-91			
181	H181DHA	183	H183DHA	185	H185DHA	187	H187EHA	189	H189EHA
182	H182DHA	184	H184DHA	186	H186EHA	188	H188EHA	191	H191EHA
200-206		Ford Transit 190	Carlyle	B20F*	1985-86 Ex Stevensons, 1994 *205 is B18F			
200	B730YUD	202	B875EOM	204	B734YUD	205	C85AUB	206	C726JJO
201	B732YUD	203	B733YUD						
212	F44XVP	Iveco Daily 40.10	Carlyle Dailybus 2	B21F	1989	Ex Carlyle demonstrator, 1989		
218-229		Freight Rover Sherpa	Carlyle Citybus 2	B20F	1990-91			
218	H708LOL	219	H709LOL	220	H710LOL	221	H731LOL	229	H729LOL
231	E231NFX	Freight Rover Sherpa	Carlyle Citybus 2	B20F	1987	Ex London & Country, 1990		
234	E234NFX	Freight Rover Sherpa	Carlyle Citybus 2	B20F	1987	Ex Shamrock & Rambler, 1989		
251	E151AJC	Freight Rover Sherpa	Carlyle Citybus 2	B20F	1988	Ex Crosville Wales, 1994		
259	D219OOJ	Freight Rover Sherpa	Carlyle	B18F	1987	Ex King Offa Travel, Westbury, 1992		
275	F275CEY	Iveco Daily 49.10	Robin Hood City Nippy	DP21F	1988	Ex Crosville Wales, 1991		
283	F483EJC	Iveco Daily 49.10	Carlyle Dailybus 2	DP25F	1989	Ex Crosville Wales, 1991		
284	F484EJC	Iveco Daily 49.10	Carlyle Dailybus 2	DP25F	1989	Ex Crosville Wales, 1991		
285	F485EJC	Iveco Daily 49.10	Carlyle Dailybus 2	DP25F	1989	Ex Crosville Wales, 1991		
286	F486EJC	Iveco Daily 49.10	Carlyle Dailybus 2	DP25F	1989	Ex Crosville Wales, 1991		
296	F276CEY	Iveco Daily 49.10	Robin Hood City Nippy	DP25F	1988	Ex Crosville Wales, 1991		

Crossing the English Bridge over the River Severn in Shrewsbury is Midlands 296, F276CEY, and one of two Robin Hood-bodied Iveco Daily 49.10s transferred from Crosville Wales. Even though the various British Bus fleets carry differing liveries vehicles have frequently moved between fleets. Midland fleet having four livery changes in fifteen years, 296 is seen in the latest scheme.
Tony Wilson

A livery which proved particularly difficult to keep clean was the dark red scheme introduced in 1990, initially intended for a selected few vehicles. Shown here is Northern Counties-bodied Renault-Dodge S56 341, E91WCM and one of ten transferred from North Western in 1991.
Barrie Kelsall

301-328			Iveco Daily 49-10		Carlyle Dailybus		B23F	1989-90		
301	F601EHA	306	F606EHA	311	F611EHA	316	F616EHA		324	F624EHA
302	F602EHA	307	F607EHA	312	F612EHA	319	F619EHA		325	F625EHA
303	F603EHA	308	F608EHA	313	F613EHA	320	F620EHA		326	F626EHA
304	F604EHA	309	F609EHA	314	F614EHA	322	F622EHA		327	G327PHA
305	F605EHA	310	F610EHA	315	F615EHA	323	F623EHA		328	G328PHA

329-339			Renault-Dodge S56		Northern Counties		B23F	1990-91		
329	H329DHA	332	H332DHA	334	H334DHA	336	H336DHA		338	H338DHA
330	H330DHA	333	H433DHA	335	H335DHA	337	H337DHA		339	H339DHA
331	H331DHA									

340-350			Renault-Dodge S56		Northern Counties		B23F	1988	Ex North Western, 1991	
340	E90WCM	343	E93WCM	345	E95WCM	347	E97WCM		349	E99WCM
341	E91WCM	344	E94WCM	346	E96WCM	348	E98WCM		350	E611LFV

351	D401NNA	Renault-Dodge S46	Northern Counties	B22F	1987	Ex Bee Line Buzz, 1992
352	D402NNA	Renault-Dodge S46	Northern Counties	B22F	1987	Ex Bee Line Buzz, 1992
353	D448NNA	Renault-Dodge S46	Northern Counties	B22F	1987	Ex Bee Line Buzz, 1992
357	D430NNA	Renault-Dodge S46	Northern Counties	B22F	1987	Ex Bee Line Buzz, 1993
358	D438NNA	Renault-Dodge S46	Northern Counties	B22F	1987	Ex Bee Line Buzz, 1993
359	D319DEF	Renault-Dodge S56	Northern Counties	B22F	1987	Ex Cleveland Transit, 1992
360	E110JPL	Renault-Dodge S56	Northern Counties	B23F	1988	Ex C-Line, 1992
361	D422NNA	Renault-Dodge S46	Northern Counties	B22F	1987	Ex Bee Line Buzz, 1992
362	D322DEF	Renault-Dodge S56	Northern Counties	B22F	1987	Ex Cleveland Transit, 1992
363	D450NNA	Renault-Dodge S46	Northern Counties	B22F	1987	Ex Bee Line Buzz, 1992
365	E325JVN	Renault-Dodge S56	Northern Counties	B22F	1987	Ex Cleveland Transit, 1992
366	D429NNA	Renault-Dodge S46	Northern Counties	B22F	1987	Ex Bee Line Buzz, 1993
367	D434NNA	Renault-Dodge S46	Northern Counties	B22F	1987	Ex Bee Line Buzz, 1993
370	D444NNA	Renault-Dodge S46	Northern Counties	B22F	1987	Ex Bee Line Buzz, 1993

The North & West Midlands Bus Handbook

371-391		Mercedes-Benz 709D		Alexander Sprint		B29F*	1995-96	*376-90 are B27F	
371	M371EFD	376	M376EFD	380	M380EFD	384	N784EHA	388	N788EHA
372	M372EFD	377	M377EFD	381	M381EFD	385	N785EHA	389	N789EHA
373	M373EFD	378	M378EFD	382	N782EHA	386	N786EHA	390	N790EHA
374	M374EFD	379	M379EFD	383	N783EHA	387	N787EHA	391	N791EHA
375	M375EFD								

400	G150GOL	Iveco Daily 49.10	Carlyle Dailybus 2	B25F	1990	Ex Carlyle, 1991

411-428		Mercedes-Benz 811D		Carlyle		B33F	1989-90 Ex C-Line, 1991-92
							414 ex Bee Line Buzz, 1993
411	G111TND	415	G115TND	421	G121TJA	426	G126TJA	428	G128TJA
414	G114TND	417	G117TND	422	G122TJA	427	G127TJA		

431-436		Mercedes-Benz 811D		LHE		B31F	1990	Ex C-Line, 1992
431	H131CDB	433	H133CDB	434	H134CDB	435	H135CDB	436	H136CDB
432	H132CDB								

442	H112DDS	Mercedes-Benz 811D	Carlyle	B33F	1990	Ex Harte Coaches, Greenock, 1995
446	H196JVT	Mercedes-Benz 814D	Wright Nim-bus	B33F	1990	Ex Stevensons, 1994
447	G897TGG	Mercedes-Benz 811D	Reeve Burgess Beaver	B33F	1990	Ex Stevensons, 1995
448	F148USX	Mercedes-Benz 811D	Alexander AM	DP33F	1988	Ex Happy Days, Woodseaves, 1991
449	G399FSF	Mercedes-Benz 811D	PMT Ami	B33F	1990	Ex Stevensons, 1994
450	G900TJA	Mercedes-Benz 811D	Mellor	B32F	1990	Ex Stevensons, 1994

451-462		Mercedes-Benz 811D		Marshall C16		B31F	1995		
451	M451EDH	454	M454EDH	457	M457EDH	459	M459EDH	461	M461EDH
452	M452EDH	455	M455EDH	458	M458EDH	460	M460EDH	462	M462EDH
453	M453EDH	456	M456EDH						

463-472		Mercedes-Benz 811D		Alexander Sprint		B31F	1995		
463	N463EHA	465	N465EHA	467	N467EHA	469	N469EHA	471	N471EHA
464	N464EHA	466	N466EHA	468	N468EHA	470	N470EHA	472	N472EHA

480	JOX480P	Leyland National 11351/1R		B47F	1976	
490	F700LCA	Mercedes-Benz 709D	Reeve Burgess Beaver	B23F	1989	Ex C-Line, 1992
493	F703LCA	Mercedes-Benz 709D	Reeve Burgess Beaver	B25F	1989	Ex C-Line, 1992
495	F705LCA	Mercedes-Benz 709D	Reeve Burgess Beaver	B25F	1989	Ex C-Line, 1992
501	H501GHA	Dennis Dart 9SDL3003	East Lancashire EL2000	B35F	1991	

1994 saw the delivery of twenty-two East Lancashire-bodied Dennis Darts which were divided between Wellington and Cannock depots. One from the former is working the Shropshire Bus tendered service between Wellington and Shrewsbury that operates through High Ercall and much rural countryside.
Richard Godfrey

HOCKLEY
via Belgrave
8
MIDLAND
371
M371 EFD

K38
MACCLESFIELD
MIDLAND
F51 ENF

502-523 Dennis Dart 9SDL3034   East Lancashire EL2000   B33F   1994

502	L502BNX	507	L507BNX	512	L512BNX	516	L516BNX	520	L620BNX
503	L503BNX	508	L508BNX	513	L513BNX	517	L517BNX	521	L521BNX
504	L504BNX	509	L509BNX	514	L514BNX	518	L618BNX	522	L522BNX
505	L605BNX	510	L510BNX	515	L515BNX	519	L519BNX	523	L523BNX
506	L506BNX	511	L511BNX						

575	NOE575R	Leyland National 11351A/1R		B49F	1977 Ex Midland Red East, 1982

647-700 Leyland National 11351A/1R   B49F*   1977-78 647 ex Midland Red East, 1982
*687/97, 700 are B47F

647	PUK647R	685	TOF685S	692	TOF692S	697	TOF697S	699	TOF699S
684	TOF684S	687	TOF687S	693	TOF693S	698	TOF698S	700	TOF700S

702-767 Leyland National 11351A/1R   B49F   1977-80 Ex Midland Red, 1981
718 is B47F

702	TOF702S	705	TOF705S	719	TOF719S	764	BVP764V	767	BVP767V
703	TOF703S	718	TOF718S	763	BVP763V	765	BVP765V		

801	J701NHA	Dennis Dart 9.8SDL3004	East Lancashire EL2000	B40F	1991	
802	M802MOJ	Dennis Dart 9.8SDL3040	Marshall C37	B40F	1994	
803	M803MOJ	Dennis Dart 9.8SDL3040	Marshall C37	B40F	1994	
804	M804MOJ	Dennis Dart 9.8SDL3054	Marshall C37	B40F	1994	
805	M805MOJ	Dennis Dart 9.8SDL3054	Marshall C37	B40F	1994	
806	N806EHA	Dennis Dart 9.8SDL3054	East Lancashire	B40F	1995	
807	N807EHA	Dennis Dart 9.8SDL3054	East Lancashire	B40F	1995	
808	N808EHA	Dennis Dart 9.8SDL3054	East Lancashire	B40F	1995	
859	TPE159S	Leyland National 11351A/1R (6HLXB)		B49F	1978	Ex Alder Valley, 1990
863	TPE163S	Leyland National 11351A/1R (6HLXB)		B49F	1978	Ex Alder Valley, 1990
866	TPE166S	Leyland National 11351A/1R (6HLXB)		B49F	1978	Ex Alder Valley, 1990
872	GMB372T	Leyland National 11351A/1R (6HLXB)		B49F	1978	Ex C-Line, 1992
873	GMB373T	Leyland National 11351A/1R (6HLXB)		B49F	1978	Ex Crosville, 1989
874	GMB374T	Leyland National 11351A/1R (6HLXB)		B49F	1978	Ex Bee Line Buzz, 1990
875	LFR875X	Leyland National 2 NL106L11/1R East Lancs Greenway (1995)	B41F	1981	Ex North Western, 1995
876	GMB376T	Leyland National 11351A/1R (6HLXB)		B49F	1978	Ex Crosville, 1989
878	GMB378T	Leyland National 11351A/1R (6HLXB)		B47F	1979	Ex Crosville, 1989
883	GMB383T	Leyland National 11351A/1R (6HLXB)		B49F	1978	Ex C-Line, 1992
890	GMB390T	Leyland National 11351A/1R (6HLXB)		B49F	1978	Ex Crosville, 1989
891	KMA401T	Leyland National 11351A/1R (6HLXB)		B47F	1979	Ex C-Line, 1992
892	KMA402T	Leyland National 11351A/1R (6HLXB)		B47F	1979	Ex C-Line, 1992
901	TOF701S	Leyland National 11351A/1R		B49F	1978	
904	TOF704S	Leyland National 11351A/1R (Cummins)		B49F	1978	
917	JOX717P	Leyland National 11351A/1R (Volvo)		B49F	1976	
937	PUK637R	Leyland National 11351A/1R	East Lancs Greenway (1994)	B49F	1977	Ex Midland Red, 1981

Midland currently operate only three Leyland National Greenway conversions, two on 11.3 metre vehicles and one based on a 10.6 mark 2 version. They lost their pods during the conversion and thus have been numbered in the 9xx series and 952, PUK652R is pictured working local commercial service 11 in Shrewsbury with the Gothic-style rail station building in the background.

939	PUK639R	Leyland National 11351A/1R (Cummins)		B49F	1977	
952	PUK652R	Leyland National 11351A/1R	East Lancs Greenway (1994)	B49F	1977	Ex Midland Red East, 1983
953	PUK653R	Leyland National 11351A/1R		B45F	1977	Ex Midland Red East, 1982
968	BVP968V	Leyland National 11351A/1R		B41F	1980	
969	BVP969V	Leyland National 11351A/1R		B49F	1980	
990	TOF690S	Leyland National 11351A/1R		B49F	1978	
1001	TR6147	Bristol LH6L	Hants & Dorset(1982)	Ch25F	1974	Ex Shamrock & Rambler, 1988

1201-1210

Dennis Falcon HC SDA421　　East Lancashire EL2000　　B48F　1990　Ex London & Country, 1991

1201	G301DPA	1207	G307DPA	1208	G308DPA	1209	G309DPA
1206	G306DPA						

(1210 G310DPA)

1211-1219

Dennis Falcon HC SDA423　　East Lancashire EL2000　　B48F　1992-93

1211	K211UHA	1213	K213UHA	1215	K215UHA	1217	K217UHA	1219	K219UHA
1212	K212UHA	1214	K214UHA	1216	K216UHA	1218	K218UHA		

1301-1305

Dennis Dart SLF　　Plaxton Pointer　　B37F　1996　*1301 is B40F

1301	N301ENX	1302	N302ENX	1303	N303ENX	1304	N304ENX	1305	N305ENX

1401	M401EFD	Scania N113CRL	East Lancashire	B42F	1995	
1402	M402EFD	Scania N113CRL	East Lancashire	B42F	1995	
1403	M403EFD	Scania N113CRL	East Lancashire	B42F	1995	
1404	M404EFD	Scania N113CRL	East Lancashire	B42F	1995	
1516	B516OEH	Leyland Tiger TRCTL11/3RH	Duple Laser 2	C53F	1985	
1522	BPR102Y	Leyland Tiger TRCTL11/3R	Duple Laser	C50F	1983	Ex London & Country, 1991
1523	123TKM	Volvo B58-56	Plaxton Supreme IV	C53F	1979	Ex Blue Bus Services, 1995
1526	BPR106Y	Leyland Tiger TRCTL11/3R	Duple Laser	C50F	1983	Ex London & Country, 1991
1527	BPR107Y	Leyland Tiger TRCTL11/3R	Duple Laser	C50F	1983	Ex London & Country, 1991
1604	B604OEH	Leyland Tiger TRCTL11/3RH	Duple Laser 2	C55F	1984	
1605	B605OEH	Leyland Tiger TRCTL11/3RH	Duple Laser 2	C55F	1984	
1606	B606OEH	Leyland Tiger TRCTL11/3RH	Duple Laser 2	C55F	1984	
1607	B607OEH	Leyland Tiger TRCTL11/3RH	Duple Laser 2	C55F	1984	
1615	A215PEV	Leyland Tiger TRCTL11/2R	Duple Dominant IV Express	DP53F	1983	Ex Southdown, 1990
1639	A139EPA	Leyland Tiger TRCTL11/2R	Plaxton Paramount 3200 E	C53F	1984	Ex C-Line, 1992
1654	TDC854X	Leyland Tiger TRCTL11/3R	Duple Dominant IV Express	C53F	1982	Ex Shamrock & Rambler, 1998
1660	A160EPA	Leyland Tiger TRCTL11/3R	Plaxton Paramount 3200 E	C50FT	1984	Ex C-Line, 1992
1698	A898KAH	Leyland Tiger TRCTL11/3RH	Plaxton Paramount 3200 E	C53F	1983	Ex C-Line, 1992

1701-1709

Leyland Tiger TRCTL11/2R　　Duple Dominant　　B51F　1984

1701	A701HVT	1703	A703HVT	1705	A705HVT	1707	A707HVT	1709 A709HVT
1702	A702HVT	1704	A704HVT	1706	A706HVT	1708	A708HVT	

1710-1720

Leyland Tiger TRCTL11/2R　　East Lancashire (1989)　　B51F*　1982　Ex London & Country, 1989
*1710/3/4/7/8 are DP49F; 1712 is B55F

1710	TPC101X	1713	TPC103X	1715	WPH125Y	1717	TPC107Y	1719 WPH139Y
1711	WPH121Y	1714	TPC104X	1716	WPH126Y	1718	TPC114Y	1720 WPH122Y
1712	TPC102X							

1721-1729

Leyland Tiger TRCTL11/3RH　　East Lancashire (1991)　　B59F　1984-86　Ex London & Country, 1991

1721	C141SPB	1723	B103KPF	1725	B105KPF	1728	B108KPF	1729 B109KPF
1722	B102KPF	1724	B104KPF	1726	C262SPC			

1730	YPJ207Y	Leyland Tiger TRCTL11/3R	East Lancashire (1992)	B59F	1982	Ex County, 1991
1733	LTS93X	Leyland Tiger TRCTL11/3R	East Lancashire (1992)	B59F	1982	Ex Tame Valley, Birmingham, 1992
1735	DJN25X	Leyland Tiger TRCTL11/2R	East Lancashire (1992)	B53F	1982	Ex County, 1992
1737	UJN430Y	Leyland Tiger TRCTL11/2R	East Lancashire (1991)	B53F	1982	Ex County, 1991
1738	WPH118Y	Leyland Tiger TRCTL11/2R	East Lancashire (1992)	B53F	1983	Ex County, 1991
1740	AAX590A	Leyland Tiger TRCTL11/3R	East Lancashire (1993)	B59F	1984	Ex Rhondda, 1992
1742	A42SMA	Leyland Tiger TRCTL11/2R	East Lancashire (1992)	B53F	1984	Ex North Western, 1991
1743	WPH123Y	Leyland Tiger TRCTL11/2R	East Lancashire (1992)	B53F	1983	Ex County, 1991

1745-1752
Leyland Tiger TRBTL11/3ARZA   Alexander N   B53F   1988   Ex Timeline, 1993-95

1745	E25UNE	1747	E27UNE	1749	E29UNE	1751	E31UNE	1752	E32UNE
1746	E26UNE	1748	E28UNE	1750	E30UNE				

1753-1772
Leyland Tiger TRBL10/3ARZA   Alexander N   B53F*   1988-89   Ex Timeline, 1994-95
*1759-77 are B55F

1753	F33ENF	1755	F35ENF	1759	F39ENF	1771	F51ENF	1772	F52ENF
1754	F34ENF	1756	F36ENF	1760	F40ENF				

1778   F278HOD   Leyland Tiger TRBTL11/2RP   Plaxton Derwent 2   B54F   1988   Ex Thames Transit, 1994

1801-1806
Dennis Dominator DDA1032   East Lancashire   H47/29F   1990   1803-6 are DDA1031

1801	G801THA	1803	H803AHA	1804	H804AHA	1805	H805AHA	1806	H806AHA
1802	G802THA								

1823	BMA523W	Bristol VRT/SL3/6LXB	Eastern Coach Works	H43/31F	1981	Ex Crosville Wales, 1991
1831	M831SDA	Scania N113DRB	East Lancashire	DPH43/29F	1995	
1832	M832SDA	Scania N113DRB	East Lancashire	DPH43/29F	1995	
1833	M833SDA	Scania N113DRB	East Lancashire	DPH43/29F	1995	
1834	M834SDA	Scania N113DRB	East Lancashire	H45/33F	1995	
1835	M835SDA	Scania N113DRB	East Lancashire	H45/33F	1995	
1858	VCA458W	Bristol VRT/SL3/6LXB	Eastern Coach Works	H43/31F	1981	Ex Crosville Wales, 1991
1860	VCA460W	Bristol VRT/SL3/6LXB	Eastern Coach Works	H43/31F	1981	Ex Crosville Wales, 1991
1870	WTU470W	Bristol VRT/SL3/6LXB	Eastern Coach Works	H43/31F	1981	Ex Crosville Wales, 1991

1903-1910
Leyland Olympian ONLXB/1R   Eastern Coach Works   H45/32F   1983

1903	EEH903Y	1906	EEH906Y	1907	EEH907Y	1909	EEH909Y	1910	EEH910Y
1905	EEH905Y								

1911	B911NBF	Leyland Olympian ONLXB/1R	Eastern Coach Works	DPH42/28F	1984	
1912	B912NBF	Leyland Olympian ONLXB/1R	Eastern Coach Works	DPH42/28F	1984	
1913	B913NBF	Leyland Olympian ONLXB/1R	Eastern Coach Works	DPH42/28F	1984	
1914	B197DTU	Leyland Olympian ONLXB/1R	Eastern Coach Works	H45/32F	1985	Ex Crosville, 1989
1915	B198DTU	Leyland Olympian ONLXB/1R	Eastern Coach Works	H45/32F	1985	Ex Crosville, 1989
1916	G916LHA	Leyland Olympian ON2R50G16ZA	East Lancashire	H45/29F	1989	
1917	G917LHA	Leyland Olympian ON2R50G16ZA	East Lancashire	H45/29F	1989	
1918	G918LHA	Leyland Olympian ON2R50G16ZA	East Lancashire	H45/29F	1989	
1919	G919LHA	Leyland Olympian ON2R50G16ZA	East Lancashire	H45/29F	1989	
1923	B203DTU	Leyland Olympian ONLXB/1R	Eastern Coach Works	DPH42/27F	1985	Ex Crosville Wales, 1990
1924	B204DTU	Leyland Olympian ONLXB/1R	Eastern Coach Works	DPH42/27F	1985	Ex Crosville Wales, 1990
1937	GFM107X	Leyland Olympian ONLXB/1R	Eastern Coach Works	H45/32F	1982	Ex Crosville, 1989
1938	PFM130Y	Leyland Olympian ONLXB/1R	Eastern Coach Works	H45/32F	1983	Ex Crosville, 1989
1950	A150UDM	Leyland Olympian ONLXB/1R	Eastern Coach Works	H45/32F	1983	Ex Stevensons, 1995
1954	A154UDM	Leyland Olympian ONLXB/1R	Eastern Coach Works	H45/32F	1984	Ex Crosville, 1989
1955	A155UDM	Leyland Olympian ONLXB/1R	Eastern Coach Works	H45/32F	1984	Ex Crosville, 1989
1972	A172VFM	Leyland Olympian ONLXB/1R	Eastern Coach Works	H45/32F	1984	Ex C-Line, 1992
2005	G505SFT	Leyland Olympian ONCL10/1RZ	Northern Counties Palatine	H47/30F	1989	Ex Bee Line Buzz, 1993
2007	G507SFT	Leyland Olympian ONCL10/1RZ	Northern Counties Palatine	H47/30F	1989	Ex Bee Line Buzz, 1993
2010	G510SFT	Leyland Olympian ONCL10/1RZ	Northern Counties Palatine	H47/30F	1989	Ex Bee Line Buzz, 1993
2011	G511SFT	Leyland Olympian ONCL10/1RZ	Northern Counties Palatine	H47/30F	1989	Ex Bee Line Buzz, 1993
2044	G644BPH	Volvo Citybus B10M-50	Northern Counties Palatine	H45/35F	1989	Ex Bee Line Buzz, 1993
2045	G645BPH	Volvo Citybus B10M-50	Northern Counties Palatine	H45/35F	1989	Ex Bee Line Buzz, 1993
2046	G646BPH	Volvo Citybus B10M-50	Northern Counties Palatine	H45/35F	1989	Ex Bee Line Buzz, 1993
2047	G647BPH	Volvo Citybus B10M-50	Northern Counties Palatine	H45/35F	1989	Ex Bee Line Buzz, 1993

Previous Registrations:

123TKM	DVO1T		LTS93X	VSS1X, WLT610
AAX590A	A217VWO		UJN430Y	WPH124Y, FBZ2514
DJN25X	TPC106X, OIB3510			

Livery: Yellow and red; yellow and blue (Park & Ride) 1301-5, 1401-4

Opposite: **Representing the Midland double-deck fleet are 1833, M833SDA, a Scania N113 with East Lancashire bodywork incorporating high-back seating, and 2010, G510SFT, a Leyland Olympian with Northern Counties bodywork. Stevensons vehicles are gradually being brought into a single fleet. So far Midland are operating four vehicles from that fleet while some of Midlands coaches are now allocated to Stephensons depots.**   *Tony Wilson*

NUNEATON X76
via Drayton Manor/Tamworth
MIDLAND
M833 SDA

BIRMINGHAM
VIA KINGSBURY 116
We never accept money from strangers.
United friendly
PERSON
TO PERSON
MIDLAND
2010
G510 SFT

Minsterley Motors' stage services are now often undertaken by the latest pair of Duple Dominant-bodied Bedfords added to the fleet during the last year. Seen leaving Shrewsbury bus station for Bishops Castle is of this pair, BGR684W. *Cliff Beeton*

Moorland Buses operate minibus service through the Potteries area using eight minibuses acquired from the Blue Buses business at Bucknall. The majority are Freight Rover Sherpa with Dormobile bodywork but two, including D62NOF seen here, carry Carlyle bodywork. The vehicle was photographed entering Newcastle bus station. *Tony Wilson*

MINSTERLEY MOTORS

JB Jones, LA & C Evans, Stiperstones, Minsterley, Shropshire SY5 0LZ

510DMY	Leyland Leopard PSU5/4R	Plaxton Elite III	C53F	1973	Ex Young, Romsley, 1987
RAX806M	Bedford YRT	Plaxton Elite Express III	C53F	1974	Ex Clun Valley Motors, 1989
ONL924M	Bedford YRT	Plaxton Elite III	C53F	1974	Ex Dore, Leafield, 1981
MFV70P	Bedford YRT	Duple Dominant	C53F	1975	Ex Holmeswood, Rufford, 1981
NWK10P	Bedford YLQ	Plaxton Supreme III	C45F	1976	Ex Phillips, Broad Oak, 1991
MWB115P	Bedford YLQ	Plaxton Supreme III	C45F	1976	Ex Torr, Gedling, 1991
RRR520R	Bedford YMT	Plaxton Supreme III Express	C53F	1977	Ex Orion Travel, Hereford, 1989
PYG139R	Bedford YMT	Plaxton Supreme III	C53F	1977	Ex Yeomans, Hereford, 1989
SNM71R	Bedford YMT	Plaxton Supreme III	C53F	1977	Ex Delro, Mytchett, 1990
CWA439T	Bedford YMT	Duple Dominant	B55F	1979	Ex Beeline, Warminster, 1993
KTA356V	Bedford YMT	Duple Dominant II Express	C53F	1980	Ex Thomas, Relubbus, 1995
MHP17V	Bedford YMT	Plaxton Supreme IV	C49F	1980	Ex Harry Shaw, Coventry, 1982
1877NT	Bedford YMT	Plaxton Supreme IV Express	C53F	1980	Ex Clun Valley Motors, 1991
BGR683W	Bedford YMT	Duple Dominant	B53F	1980	Ex Jolly, South Hylton, 1995
BGR684W	Bedford YMT	Duple Dominant	B53F	1980	Ex Jolly, South Hylton, 1995
JUS476Y	Mercedes-Benz L608D	Whittaker	C19F	1983	Ex Pasadena Roof Orchestra, 1990
WJH503Y	Bedford YNT	Plaxton Paramount 3500	C53F	1983	Ex Smith, Liss, 1991
A615KRT	Bedford YNT	Plaxton Paramount 3200	C53F	1984	Ex Ward, Alresford, 1990
F999PLA	Volkswagen LT55	Optare City Pacer	B25F	1988	Ex Clapton Coaches, 1994
G738VKK	Renault Master T35D	Jubilee	M16	1990	Ex Van, 1992

Previous Registrations:
1877NT   JRW767V   A615KRT   A253SBM, 866VNU
510DMY   KNR327L, 961CUF, FDH926L   WJH503Y   9489PH

Livery: Grey and blue

MOORLAND BUSES

S A Titterton, Weston Service Station, Weston Coyney, Stafford ST3 6QB

C468TAY	Ford Transit 160	Rootes	B16F	1985	Ex Blue Buses, Bucknall, 1995
C573TUT	Ford Transit 160	Dormobile	B16F	1986	Ex Blue Buses, Bucknall, 1995
D598VBV	Freight Rover Sherpa	Dormobile	B16F	1986	Ex Blue Buses, Bucknall, 1995
D788JUB	Freight Rover Sherpa	Dormobile	B20F	1986	Ex Blue Buses, Bucknall, 1995
D167NON	Freight Rover Sherpa	Carlyle	B16F	1987	Ex Blue Buses, Bucknall, 1995
D62NOF	Freight Rover Sherpa	Carlyle	B16F	1987	Ex Blue Buses, Bucknall, 1995
D39TKA	Freight Rover Sherpa	Dormobile	B16F	1987	Ex Blue Buses, Bucknall, 1995
E969SOF	Freight Rover Sherpa	Dormobile	B20F	1987	Ex Blue Buses, Bucknall, 1995

Livery: Blue and white

NCB

NCB Motors Ltd, Edstaston Garage, Wem, Shropshire SY4 6RF

MRY53P	Bedford YMT	Duple Dominant	C53F	1976	Ex Glennie, New Mill, 1989
PDK308S	Bedford YLQ	Duple Dominant	C45F	1977	Ex Garratt, Leicester, 1990
KDM760T	Bedford YMT	Duple Dominant II	C53F	1979	Ex Fraser, Rufford, 1985
A32GJT	Bedford YNT	Duple Laser	C53F	1984	Ex Marchwood, Totton, 1987
A33GJT	Bedford YNT	Duple Laser	C53F	1984	Ex Marchwood, Totton, 1987
B977HNT	Bedford YNV Venturer	Duple Laser 2	C55F	1985	
F374DUX	Volvo B10M-60	Duple 340	C55F	1988	
F971HGE	Volvo B10M-60	Plaxton Paramount 3500 III	C57F	1989	Ex Park's, 1990
J1NCB	Volvo B10M-60	Jonckheere Deauville P599	C51FT	1991	
K1NCB	Volvo B10M-60	Jonckheere Deauville P599	C55FT	1992	
L1NCB	Volvo B10M-60	Plaxton Premiére 350	C53FT	1994	

Livery: Cream and brown

NCB Motors have gradually displaced Bedford coaches with Volvo, most of which carry appropriate Select index marks. The only second-hand Volvo is F971HGE which is fitted with a Plaxton Paramount 3500 mark III body. It is seen in Victoria in the company's attractive cream and brown livery. *Colin Lloyd*

NCP

National Car Parks Ltd, International Airport, Birmingham   B26 3QZ

A386NNK	Mercedes-Benz L608D	Reeve Burgess	DP16F	1983	Ex Capital, West Drayton, 1993
E624FLD	Mercedes-Benz 609D	Reeve Burgess Beaver	B16F	1988	Ex Capital, West Drayton, 1993
E626FLD	Mercedes-Benz 609D	Reeve Burgess Beaver	B16F	1988	Ex Capital, West Drayton, 1993
E627FLD	Mercedes-Benz 609D	Reeve Burgess Beaver	B16F	1988	Ex Capital, West Drayton, 1993
M442BLC	Dennis Dart 9SDL3051	Plaxton Pointer	B25D	1995	
M443BLC	Dennis Dart 9SDL3051	Plaxton Pointer	B25D	1995	
M445BLC	Dennis Dart 9SDL3051	Plaxton Pointer	B25D	1995	
M446BLC	Dennis Dart 9SDL3051	Plaxton Pointer	B25D	1995	

Livery: Yellow

National Car Parks operate courtesy services between Birmingham airport terminal and the various car parks for the airport. Seen outside the newer Eurohub terminal used by British Airways is M442BLC, one of the Dennis Darts with Plaxton Pointer bodywork equipped with extra luggage space employed on the service. *Bill Potter*

NORTH BIRMINGHAM

North Birmingham Busways Ltd, 38 Wood Lane, Erdington, West Midlands   B24 9QN

17	JFV317S	Leyland Atlantean AN68A/2R	East Lancashire	H50/36F	1978	Ex Blackpool, 1994
18	JFV318S	Leyland Atlantean AN68A/2R	East Lancashire	H50/36F	1978	Ex Blackpool, 1994
19	JFV319S	Leyland Atlantean AN68A/2R	East Lancashire	H50/36F	1978	Ex Blackpool, 1994
20	JFV320S	Leyland Atlantean AN68A/2R	East Lancashire	H50/36F	1978	Ex Blackpool, 1994
21	URN321V	Leyland Atlantean AN68A/2R	East Lancashire	H50/36F	1979	Ex Blackpool, 1994
24	STK124T	Leyland Atlantean AN68A/1R	Roe	H43/28F	1979	Ex Southampton Citybus, 1995
25	STK125T	Leyland Atlantean AN68A/1R	Roe	H43/28F	1979	Ex Southampton Citybus, 1995
29	STK129T	Leyland Atlantean AN68A/1R	Roe	H43/31F	1979	Ex Plymouth Citybus, 1994
30	STK130T	Leyland Atlantean AN68A/1R	Roe	H43/31F	1979	Ex Plymouth Citybus, 1995
31	STK131T	Leyland Atlantean AN68A/1R	Roe	H43/31F	1979	Ex Plymouth Citybus, 1996

Livery: Cream and green

Since the last edition of the book North Birmingham Busways have increased their fleet with the purchase of five Roe-bodied Atlanteans from the same original batch. All are now converted to single doorway and carry green and cream livery to the Blackpool layout. Seen near the Bull-Ring is 25, STK125T. *Tony Wilson*

OWEN'S

F G Owen, 32/34 Beatrice Street, Oswestry, Shropshire   SY11 1QG

Depot : Red Lion Garage, Horsemarket, Oswestry

w	BXI2410	Bova FLD12.280	Bova Futura	C53F	1984	Ex Boulton, Cardington, 1994
	B216GUX	Renault Trafic	Holdsworth	M8	1985	Ex Tanat Valley, Pentrefellin, 1995
	E880YNT	Renault Trafic	Holdsworth	M11	1988	Ex private owner, 1995
	F258BHF	Volvo B10M-60	Plaxton Expressliner	C46FT	1990	Ex Arvonia, Llanrug, 1995
	G373REG	Volvo B10M-60	Plaxton Expressliner	C49FT	1990	Ex Premier Travel, 1996
	G106AVX	Dennis Javelin 12SDA1912	Duple 320	C52FT	1990	Ex Colchester, 1993
	H237RUX	Hestair Duple SDA1512	Duple 425	C51FT	1991	
	K879EAW	Volvo B10M-60	Plaxton Premiére 350	C49FT	1993	
	K106UFP	Toyota Coaster HDB30R	Caetano Optimo II	C21F	1993	
	L64YJF	Toyota Coaster HZB50R	Caetano Optimo III	C21F	1993	
	L776LUJ	Volvo B10M-60	Plaxton Excalibur	C53F	1994	
	M954HRY	Dennis Javelin 12SDA2136	Caetano Algarve II	C55F	1994	
	M75OUX	Dennis Javelin 12SDA2131	Plaxton Premiére 320	C53F	1994	

Previous Registrations:
BXI2410   A118DUY, PNR723   F258BHF   NXI9006

Livery: White, red, and blue

Now the oldest vehicle among Owen's modern fleet, and recently taken out of service, is Bova Futura BXI2410 photographed in the town after its acquisition from Shropshire neighbour, Boulton. The Bova has gradually increased in popularity and is currently sold through the Optare group who are its importers. *Ralph Stevens*

PATTERSON

D F & P M Patterson, Unit 36, Elliott Road, Selly Oak, Birmingham   B29 6LR

MWJ730W	Ford R1114	Plaxton Supreme IV	C53F	1981	Ex Pickin Bingham, 1995
C457AHY	Leyland Cub CU435	Reeve Burgess	B20FL	1986	Ex County of Avon, 1994
C337CHT	Leyland Cub CU435	Reeve Burgess	B20FL	1986	Ex County of Avon, 1995
	Leyland Cub CU435	Reeve Burgess	B20DL*	1986	*Seating varies
D314EFK	D318EFK	D692FDH	D695FDH		D698FDH
D315EFK	D319EFK	D693FDH	D696FDH		D699FDH
D316EFK		D691FDH	D694FDH	D697FDH	D700FDH
D317EFK					
D627BCK	Iveco Daily 49.10	Robin Hood City Nippy	B25F	1987	Ex Ribble, 1993
D305JJD	Mercedes-Benz ?07D	Reeve Burgess	M16	1987	Ex ?, 1995
D944BAB	Mercedes-Benz 609D	PMT	C21F	1987	Ex ?, 1996
D701GHT	Leyland Cub CU435	Wadham Stringer Vanguard	B16FL	1987	Ex County of Avon, 1995
D705GHT	Leyland Cub CU435	Wadham Stringer Vanguard	B12FL	1987	Ex County of Avon, 1995
E312HLN	Renault G08	Reeve Burgess	DP15L	1988	Ex LB Islington, 1995
E76PEE	Mercedes-Benz 811D	Coachcraft	C25F	1988	Ex Robin Hood, Rudyard, 1996
E533UOK	Mercedes-Benz 609D	Reeve Burgess Beaver	C23F	1988	
E52MTC	Mercedes-Benz 609D	Reeve Burgess Beaver	C23F	1988	Ex Aztec, Bristol, 1995
E160NEU	Mercedes-Benz 609D	Reeve Burgess Beaver	C23F	1988	Ex Aztec, Bristol, 1995
F574YSC	Mercedes-Benz 609D	Alexander Sprint	C25F	1989	Ex Goosecroft, Stirling, 1996
F848RHY	Mercedes-Benz 407D	Reeve Burgess	M15	1989	Ex Aztec, Bristol, 1995
F203RAE	Mercedes-Benz 507D	Reeve Burgess	M16	1988	Ex Aztec, Bristol, 1995
F929GGE	Mercedes-Benz 609D	Made-to-Measure	C24F	1988	Ex McGovern, Newton Mearns, 1994
F968XWM	Mercedes-Benz 609D	Advanced Vehicle Bodies	C20F	1988	Ex Coachmaster, Coulsdon, 1995
F851RHY	Mercedes-Benz 609D	Reeve Burgess Beaver	B20F	1989	Ex West Midlands SNT, 1995
G141LRM	Mercedes-Benz 609D	Reeve Burgess Beaver	B20F	1989	Ex North Western, 1995
G147LRM	Mercedes-Benz 609D	Reeve Burgess Beaver	B20F	1989	Ex North Western, 1995
H557XNN	Iveco Daily 49.10	Carlyle Dailybus	B25F	1990	Ex Trent, 1993

Previous Registrations:
MWJ730W   KKW363W, 260ERY

Livery: Green, white and orange.

Recently, Pattersons have withdrawn from commercial service in preference to tendered operations mostly involving the disabled. Many of the fleet of Mercedes-Benz minibuses have now been released for service elsewhere including H741TWB, seen here towards the end of commercial services.
Bill Potter

PMT

PMT Ltd, Hobson Street, Burslem, Stoke-on-Trent ST6 2AQ

Depots: Dividy Road, Adderley Green; Scotia Road, Burslem; Brookhouse Industrial Estate, Cheadle; Liverpool Road, Chester; Second Avenue, Crewe Gates Farm, Crewe; Platt Street, Dukinfield; Bus Station, Ellesmere Port; Pasture Road, Moreton; Liverpool Road, Newcastle-under-Lyme and New Chester Road, Rock Ferry.

MXU22	H202JHP	Peugeot-Talbot Pullman	Talbot	B22F	1990	Ex Midland Red West, 1995
MXU23	H203JHP	Peugeot-Talbot Pullman	Talbot	B22F	1990	Ex Midland Red West, 1995
STL24	ERF24Y	Leyland Tiger TRCTL11/3R	Plaxton Paramount 3500	C53F	1983	
MBU25	M25YRE	Peugeot Boxer	TBP	M9	1995	
MBU26	M26YRE	Peugeot Boxer	TBP	M9	1995	
MBU27	M27YRE	Peugeot Boxer	TBP	M9	1995	
MBU28	M28YRE	Peugeot Boxer	TBP	M9	1995	
MRE29	C477EUA	Renault Master T35	Renault	M9	1986	Ex WYM Ambulance, 1996
MRE30	D810NWW	Renault Master T35	Renault	M9	1987	Ex WYM Ambulance, 1996
STL44	E44JRF	Leyland Tiger TRCTL11/3R	Plaxton Paramount 3500 III	C53F	1988	
MMM50	B232AFV	Mercedes-Benz L307D	Reeve Burgess	M12	1985	Ex Landliner, Birkenhead, 1990
SLL80	NED433W	Leyland Leopard PSU3E/4R	Plaxton Supreme IV	C53F	1981	Ex Turner, Brown Edge, 1988
MMM88	C108SFP	Mercedes-Benz L307D	Reeve Burgess	M12	1985	Ex Goldcrest, Birkenhead, 1990
MMM89	F660EBU	Mercedes-Benz 609D	North West Coach Sales	B19F	1988	Ex Landliner, Birkenhead, 1990
MMM97	D176VRP	Mercedes-Benz L608D	Alexander	B20F	1986	Ex Milton Keynes Citybus, 1992
MMM98	D185VRP	Mercedes-Benz L608D	Alexander	B20F	1986	Ex Milton Keynes Citybus, 1992
MMM99	C683LGE	Mercedes-Benz L608D	Reeve Burgess	B20F	1985	Ex Strathclyde, 1991
MMM100	F100UEH	Mercedes-Benz 609D	PMT	C24F	1989	
MMM101	G101EVT	Mercedes-Benz 609D	PMT	C21F	1990	
MMM102	F452YHF	Mercedes-Benz 609D	North West Coach Sales	C24F	1989	Ex C & M, Aintree, 1992
MMM104	F713OFH	Mercedes-Benz 307D	North West Coach Sales	M12	1989	Ex van, 1992
IFF105	J328RVT	Iveco Daily 49.10	Reeve Burgess Beaver	C29F	1991	Ex Roseville Taxis, Newcastle, 1993
MRP106	E106LVT	Renault-Dodge S56	PMT	C22F	1988	
MRP108	D162LTA	Renault-Dodge S56	Reeve Burgess	B23F	1987	Ex Cardiff Bus, 1994
MMM109	F217OFB	Mercedes-Benz 307D	North West Coach Sales	M12L	1989	Ex van, 1992
IMM110	H189CNS	Mercedes-Benz 814D	Dormobile Routemaker	C33F	1991	Ex Executive Travel, 1994
MMM112	XRF2X	Mercedes-Benz L307D	Reeve Burgess	M12	1983	
MMM114	G805AAD	Mercedes-Benz 308	North West Coach Sales	M12L	1989	Ex van, 1992
MMM115	XRF1X	Mercedes-Benz L608D	PMT Hanbridge	C21FL	1984	
MMM116	FXI8653	Mercedes-Benz L608D	PMT Hanbridge	C21FL	1984	
MMM117	B117OBF	Mercedes-Benz L608D	PMT Hanbridge	B19F	1984	

The number of full-size coaches in the PMT fleet is gradually reducing. Recently photographed at Hanley is STL43, E43JRF which has subsequently been taken out of service, its replacement is one of the new Dennis saloons.
Cliff Beeton

71

MMM120-159   Mercedes-Benz L608D   PMT Hanbridge   B20F   1985-86

120	C120VBF	128	C128VRE	137	C137VRE	145	C145WRE	153	D153BEH
121	C121VRE	130	C130VRE	138	C138VRE	146	C146WRE	154	D154BEH
122	C122VRE	131	C131VRE	139	C139VRE	147	C147WRE	155	D155BEH
123	C123VRE	132	C132VRE	140	C140VRE	148	C148WRE	156	D156BEH
124	C124VRE	133	C133VRE	141	C141VRE	149	C149WRE	157	D157BEH
125	C125VRE	134	C134VRE	142	C142VRE	150	C150WRE	158	D158BEH
126	C126VRE	135	C135VRE	143	C143VRE	151	C151WRE	159	D159BEH
127	C127VRE	136	C136VRE	144	C144VRE	152	D152BEH		

MFF178	F166DNT	Ford Transit VE6	Dormobile	M15	1989	Ex Shropshire CC, 1994

MMM182	D182BEH	Mercedes-Benz L608D	PMT Hanbridge	B20F	1986	
MMM183	D183BEH	Mercedes-Benz L608D	PMT Hanbridge	B20F	1986	
MMM185	D185BEH	Mercedes-Benz L608D	PMT Hanbridge	B20F	1986	
MMM188	D188BEH	Mercedes-Benz L608D	PMT Hanbridge	B20F	1986	
MMM190	C124LHS	Mercedes-Benz L608D	Reeve Burgess	B20F	1986	Ex Strathclyde, 1991
MMM193	D120PGA	Mercedes-Benz L608D	PMT Hanbridge	B19F	1986	Ex Strathclyde, 1991
MMM194	D121PGA	Mercedes-Benz L608D	PMT Hanbridge	B19F	1986	Ex Strathclyde, 1991
MMM195	D122PGA	Mercedes-Benz L608D	PMT Hanbridge	B19F	1986	Ex Strathclyde, 1991
MMM196	D123PGA	Mercedes-Benz L608D	PMT Hanbridge	B19F	1986	Ex Strathclyde, 1991

MMM199-209   Mercedes-Benz L608D   Alexander   B20F*   1986   Ex Milton Keynes Citybus, 1992
*205 is B19F

199	D159VRP	203	D179VRP	205	D186VRP	207	D157VRP	209	D184VRP
200	D160VRP								

MMM210	C706JMB	Mercedes-Benz L608D	Reeve Burgess	B19F	1986	Ex Crosville, 1990
MMM213	C711JMB	Mercedes-Benz L608D	Reeve Burgess	B19F	1986	Ex Crosville, 1990
MMM215	D548FAE	Mercedes-Benz L608D	Dormobile	B20F	1986	Ex City Line, 1994
MMM216	D549FAE	Mercedes-Benz L608D	Dormobile	B20F	1986	Ex City Line, 1994
MMM217	D550FAE	Mercedes-Benz L608D	Dormobile	B20F	1986	Ex City Line, 1994

MPC224-230   MCW MetroRider MF150/118   MCW   B25F   1988   Ex Crosville, 1990

224	F88CWG	226	F106CWG	228	F108CWG	229	F109CWG	230	F110CWG
225	F95CWG	227	F107CWG						

MPC231	L231NRE	Optare MetroRider	Optare	B31F	1994	

SNL287	SFA287R	Leyland National 11351A/1R		B52F	1977	
SNG298	GMB377T	Leyland National 11351A/1R		B49F	1978	Ex Crosville, 1990

For a while PMT produced bodywork for its own use and for sale to other operators. The Knype was designed for the Leyland Swift and featured a rather square outline. Shown here is 318, G318YVT, the last of the type built and which features high-back seating.
Cliff Beeton

IWC310-318
Leyland Swift LBM6T/2RS — PMT Knype — DP37F* — 1988-89 *315-8 are DP35F

310	F310REH	312	F312REH	315	F315REH	317	F317REH	318	G318YVT
311	F311REH	313	F313REH	316	F316REH				

IWC320	E342NFA	Leyland Swift LBM6T/2RS	PMT Knype	DP37F	1988	Ex PMT demonstrator, 1988
IPC321	L321HRE	Optare MetroRider	Optare	DP30F	1993	
IPC322	L269GBU	Optare MetroRider	Optare	B29F	1993	
IPC323	L323NRF	Optare MetroRider	Optare	B29F	1994	

IMM330-353
Mercedes-Benz 811D — PMT Ami — B28F — 1989-90

330	G330XRE	335	G335XRE	340	G340XRE	345	G345CBF	350	G550ERF
331	G331XRE	336	G336XRE	341	G341XRE	346	G346CBF	351	H351HRF
332	G332XRE	337	G337XRE	342	G342CBF	347	G347ERF	352	H352HRF
333	G333XRE	338	G338XRE	343	G343CBF	348	G348ERF	353	H353HRF
334	G334XRE	339	G339XRE	344	G344CBF	349	G349ERF		

IMM354	H354HVT	Mercedes-Benz 811D	Reeve Burgess Beaver	B31F	1990
IMM355	H355HVT	Mercedes-Benz 811D	Reeve Burgess Beaver	B31F	1990
IMM356	H356HVT	Mercedes-Benz 811D	Reeve Burgess Beaver	B33F	1990
IMM357	H357HVT	Mercedes-Benz 811D	Reeve Burgess Beaver	B33F	1990

IMM358-363
Mercedes-Benz 811D — PMT Ami — B29F — 1990

358	H358JRE	360	H160JRE	361	H361JRE	362	H362JRE	363	H363JRE
359	H359JRE								

MMM364	E39KRE	Mercedes-Benz L811D	PMT	B25F	1988	Ex van, 1990
IMM365	G495FFA	Mercedes-Benz 811D	PMT Ami	B28F	1990	

IMM366-371
Mercedes-Benz 811D — PMT Ami — B29F — 1991

366	H366LFA	368	H368LFA	369	H369LFA	370	H370LFA	371	H371LFA
367	H367LFA								

IMM372	H372MEH	Mercedes-Benz 811D	Whittaker-Europa	B31F	1991	
IMM373	H373MVT	Mercedes-Benz 811D	PMT Ami	B29F	1991	
IMM374	K374BRE	Mercedes-Benz 811D	Autobus Classique	B29F	1992	
IMM375	K375BRE	Mercedes-Benz 811D	Autobus Classique	B29F	1992	
IMM376	J920HGD	Mercedes-Benz 709D	Dormobile Routemaker	B29F	1991	Ex Stonier, 1994

IPC377-383
Optare MetroRider — Optare — B29F — 1994

377	M377SRE	379	M379SRE	381	M381SRE	382	M382SRE	383	M383SRE
378	M378SRE	380	M380SRE						

Optare MetroRiders are popular with many operators, especially as the product is available in different widths and lengths depending on need. Seen heading for Meir Park, still with its Badger motif in place, is IPC381, M381SRE.
Cliff Beeton

MMM405-429 Mercedes-Benz 709D Plaxton Beaver B24F 1996

405	N405HVT	410	N410HVT	415	N415HVT	420		425	
406	N406HVT	411	N411HVT	416	N416HVT	421		426	
407	N407HVT	412	N412HVT	417	N417HVT	422		427	
408	N408HVT	413	N413HVT	418	N418HVT	423		428	
409	N409HVT	414	N414HVT	419	N419HVT	424		429	

MMM430-448 Mercedes-Benz 709D Plaxton Beaver B24F 1992

430	J430WFA	434	K434XRF	438	K438XRF	442	K442XRF	446	K446XRF
431	J431WFA	435	K435XRF	439	K439XRF	443	K443XRF	447	K447XRF
432	K432XRF	436	K436XRF	440	K440XRF	444	K544XRF	448	K448XRF
433	K433XRF	437	K437XRF	441	K441XRF	445	K445XRF		

MXU449	K449XRF	Peugeot-Talbot Pullman	TBP	B18F	1992	
MMM450	E791CCA	Mercedes-Benz L609D	PMT	B20F	1988	Ex Roberts, Bootle, 1992

MMM451-466 Mercedes-Benz L609D PMT Hanbridge B20F 1987-88

451	D451ERE	454	D454ERE	457	D457ERE	460	E760HBF	464	E764HBF
452	D452ERE	455	D455ERE	458	D458ERE	461	E761HBF	465	E765HBF
453	D453ERE	456	D456ERE	459	D459ERE	462	E762HBF	466	E766HBF

MMM467	E767HBF	Mercedes-Benz 709D	PMT	B21F	1988	
MMM468	E768HBF	Mercedes-Benz 609D	PMT	B20F	1987	
MMM469	E769HBF	Mercedes-Benz 609D	Reeve Burgess	B20F	1987	
MMM470	E470MVT	Mercedes-Benz 709D	PMT	B23F	1988	
MMM471	E471MVT	Mercedes-Benz 609D	PMT	B20F	1988	
MMM472	F472RBF	Mercedes-Benz 609D	PMT	B20F	1988	
MMM473	F473RBF	Mercedes-Benz 609D	PMT	B20F	1988	
MMM474	E41JRF	Mercedes-Benz 709D	PMT	B23F	1988	
MMM475	F475VEH	Mercedes-Benz 609D	PMT	B20F	1989	
MMM476	E831ETY	Mercedes-Benz 609D	Reeve Burgess Beaver	B20F	1988	Ex Vasey, Ashington, 1990
MMM477	G477ERF	Mercedes-Benz 609D	PMT	B20F	1990	
MMM478	G478ERF	Mercedes-Benz 609D	PMT	B20F	1990	
MMM479	E384XCA	Mercedes-Benz 609D	PMT	B24F	1987	Ex Dennis's, Ashton, 1990
MMM480	H180JRE	Mercedes-Benz 709D	PMT	B20F	1990	
MMM481	H481JRE	Mercedes-Benz 709D	PMT	B25F	1990	
MMM482	H482JRE	Mercedes-Benz 609D	Whittaker Europa	B20F	1990	
MMM483	H483JRE	Mercedes-Benz 609D	Whittaker Europa	B20F	1990	
MMM484	J484PVT	Mercedes-Benz 709D	PMT	B25F	1991	
MMM485	J485PVT	Mercedes-Benz 709D	Whittaker (PMT)	B25F	1992	
MMM486	J486PVT	Mercedes-Benz 709D	Whittaker (PMT)	B25F	1992	

MMM487-498 Mercedes-Benz 709D Dormobile Routemaker B24F* 1993 *488/9 are B27F

487	K487CVT	490	K490CVT	493	L493HRE	495	L495HRE	497	L497HRE
488	K488CVT	491	K491CVT	494	L494HRE	496	L496HRE	498	L498HRE
489	K489CVT	492	K492CVT						

MRP501	E801HBF	Renault-Dodge S56	PMT	B25F	1987	

MRP502-528 Renault-Dodge S56 Alexander AM B20F* 1987 *527/8 are B25F

502	E802HBF	508	E808HBF	513	E813HBF	518	E818HBF	524	E824HBF
503	E803HBF	509	E809HBF	514	E814HBF	519	E819HBF	525	E825HBF
504	E804HBF	510	E810HBF	515	E815HBF	520	E820HBF	526	E826HBF
505	E805HBF	511	E811HBF	516	E816HBF	521	E821HBF	527	E527JRE
506	E806HBF	512	E812HBF	517	E817HBF	522	E822HBF	528	E528JRE
507	E807HBF								

MRP530	E526NEH	Renault-Dodge S56	PMT	B25F	1988	
MRP531	F531UVT	Renault-Dodge S56	PMT	B25F	1989	
MRP532	G532CVT	Renault-Dodge S56	PMT	B25F	1990	

Opposite: **Double-deck buses with PMT are supplied to Crosville, Pennine and PMT duties. Purchased from City Line in Bristol for work with Pennine fleet were two dual-doored versions of the VR. Now with PMT is DVG624, AHU515V. Currently one of the pair is based at Crewe. Many of the double-deck duties are met with standard NBC-ordered Leyland Olympians with Eastern Coach Works bodies, many arriving from the Crosville operation. Shown here is 733, A733GFA.** *Cliff Beeton*

169 ENDON
PMT
DVG624
AHU 515V

23 Hanley Tunstall Kidsgrove TALKE
A NEW MORTGAGE.
NO PORKIES.
PMT
DOG 733
Leyland
A733 GFA

MMM550	H834GLD	Mercedes-Benz 609D	North West Coach Sales	B19F	1990	Ex Capital, West Drayton, 1994		
MMM551	H835GLD	Mercedes-Benz 609D	North West Coach Sales	B19F	1990	Ex Capital, West Drayton, 1994		
MMM552	H836GLD	Mercedes-Benz 609D	North West Coach Sales	B19F	1990	Ex Capital, West Drayton, 1994		

MMM553-563

Mercedes-Benz 709D   Marshall C19   B24F   1994

553	L553LVT	556	L556LVT	558	L558LVT	560	M660SRE	562	M562SRE
554	L554LVT	557	L557LVT	559	M559SRE	561	M561SRE	563	M563SRE
555	L455LVT								

MMM564-573

Mercedes-Benz 709D   Plaxton Beaver   B24F   1994

564	M564SRE	566	M566SRE	568	M568SRE	570	M570SRE	572	M572SRE
565	M565SRE	567	M567SRE	569	M569SRE	571	M571SRE	573	M573SRE

MMM574-594

Mercedes-Benz 709D   Plaxton Beaver   B22F   1995

574	N574CEH	579	N579CEH	583	N583CEH	587	N587CEH	591	N591CEH
575	N575CEH	580	N580CEH	584	N584CEH	588	N588CEH	592	N592CEH
576	N576CEH	581	N581CEH	585	N585CEH	588	N589CEH	593	N593CEH
577	N577CEH	582	N582CEH	586	N586CEH	590	N590CEH	594	N594CEH
578	N578CEH								

DVG607	UDM450V	Bristol VRT/SL3/501(6LXB)	Eastern Coach Works	H43/31F	1980	Ex Crosville, 1990
DVG608	VCA452W	Bristol VRT/SL3/501(6LXB)	Eastern Coach Works	H43/31F	1980	Ex Crosville, 1990
DVG609	VCA464W	Bristol VRT/SL3/501(6LXB)	Eastern Coach Works	H43/31F	1980	Ex Crosville, 1990
DVG610	WTU465W	Bristol VRT/SL3/501(6LXB)	Eastern Coach Works	H43/31F	1980	Ex Crosville, 1990
DVG611	WTU472W	Bristol VRT/SL3/501(6LXB)	Eastern Coach Works	H43/31F	1980	Ex Crosville, 1990
DVG612	WTU481W	Bristol VRT/SL3/501(6LXB)	Eastern Coach Works	H43/31F	1981	Ex Crosville, 1990
DVG613	WTU482W	Bristol VRT/SL3/501(6LXB)	Eastern Coach Works	H43/31F	1981	Ex Crosville, 1990
DVG614	WTU483W	Bristol VRT/SL3/501(6LXB)	Eastern Coach Works	H43/31F	1981	Ex Crosville, 1990
DVG615	DCA526X	Bristol VRT/SL3/501(6LXB)	Eastern Coach Works	H43/31F	1981	Ex Crosville, 1990

DVG616-621

Bristol VRT/SL3/6LXB   Eastern Coach Works   H43/31F   1979   Ex Thames Transit, 1989

616	YBW487V	618	YBW489V	619	EJO490V	620	EJO491V	621	EJO492V

DVG622	507EXA	Bristol VRT/SL2/6G	Eastern Coach Works	O43/31F	1974	
DVG624	AHU515V	Bristol VRT/SL3/6LXB	Eastern Coach Works	H43/27D	1980	Ex City Line, 1994
DVG625	AHW203V	Bristol VRT/SL3/6LXB	Eastern Coach Works	H43/27D	1980	Ex City Line, 1994
DVL635	UMB333R	Bristol VRT/SL3/501	Eastern Coach Works	H43/31F	1977	Ex Crosville, 1990
DVL640	ODM409V	Bristol VRT/SL3/501	Eastern Coach Works	H43/31F	1979	Ex Crosville, 1990
DVL641	PCA420V	Bristol VRT/SL3/501	Eastern Coach Works	H43/31F	1979	Ex Crosville, 1990
DVL642	PCA421V	Bristol VRT/SL3/501	Eastern Coach Works	H43/31F	1979	Ex Crosville, 1990
DVL643	RLG430V	Bristol VRT/SL3/501	Eastern Coach Works	H43/31F	1980	Ex Crosville, 1990
DVL644	RMA443V	Bristol VRT/SL3/501	Eastern Coach Works	H43/31F	1980	Ex Crosville, 1990
DVL645	WTU488W	Bristol VRT/SL3/501	Eastern Coach Works	H43/31F	1981	Ex Crosville, 1990
DVL646	WTU489W	Bristol VRT/SL3/501	Eastern Coach Works	H43/31F	1981	Ex Crosville, 1990
DVL647	WTU491W	Bristol VRT/SL3/501	Eastern Coach Works	H43/31F	1981	Ex Crosville, 1990
DVL674	URF674S	Bristol VRT/SL3/501	Eastern Coach Works	H43/31F	1978	
DVL685	YBF685S	Bristol VRT/SL3/501	Eastern Coach Works	H43/31F	1978	
DVL689	BRF689T	Bristol VRT/SL3/501	Eastern Coach Works	H43/31F	1978	
DVL693	BRF693T	Bristol VRT/SL3/501	Eastern Coach Works	H43/31F	1978	

DVL701-732

Bristol VRT/SL3/501   Eastern Coach Works   H43/31F*   1979-80   *706/11/8/20/3/5/7-9/32 are DPH39/28F

701	GRF701V	709	GRF709V	715	GRF715V	720	MFA720V	728	NEH728W
704	GRF704V	710	GRF710V	716	GRF716V	723	MFA723V	729	NEH729W
706	GRF706V	711	GRF711V	717	MFA717V	725	NEH725W	731	NEH731W
707	GRF707V	714	GRF714V	718	MFA718V	727	NEH727W	732	NEH732W
708	GRF708V								

DOG733-747

Leyland Olympian ONLXB/1R   Eastern Coach Works   H45/32F   1983-84

733	A733GFA	736	A736GFA	739	A739GFA	742	A742GFA	745	A745JRE
734	A734GFA	737	A737GFA	740	A740GFA	743	A743JRE	746	A746JRE
735	A735GFA	738	A738GFA	741	A741GFA	744	A744JRE	747	A747JRE

DOG748	EWY78Y	Leyland Olympian ONLXB/1R	Roe	H47/29F	1983	Ex Turner, Brown Edge, 1988
DOG749	EWY79Y	Leyland Olympian ONLXB/1R	Roe	H47/29F	1983	Ex Turner, Brown Edge, 1988
DOG750	GFM101X	Leyland Olympian ONLXB/1R	Eastern Coach Works	H45/32F	1982	Ex Crosville, 1990

Comparative trials between the Leyland Lynx and the DAF SB220 with Optare Delta bodywork commenced in 1990. No additional examples of either type were ordered, though subsequently six Lynx have been acquired from Topp-Line and Westbus of Ashford. Pictured while working the 23 is PMT's first Delta, SAD801, H801GRE. *Cliff Beeton*

| DOG751 | GFM102X | Leyland Olympian ONLXB/1R | | Eastern Coach Works | H45/32F | 1982 | Ex Crosville, 1990 | |
| DOG752 | GFM103X | Leyland Olympian ONLXB/1R | | Eastern Coach Works | H45/32F | 1982 | Ex Crosville, 1990 | |

DOC753-762   Leyland Olympian ONCL11/1RZ Leyland   H47/29F*   1989   *756-762 are DPH43/29F

| 753 | G753XRE | 755 | G755XRE | 757 | G757XRE | 759 | G759XRE | 761 | G761XRE |
| 754 | G754XRE | 756 | G756XRE | 758 | G758XRE | 760 | G760XRE | 762 | G762XRE |

DOG763-783   Leyland Olympian ONLXB/1R   Eastern Coach Works   H45/32F   1982-83 Ex Crosville, 1990

763	GFM104X	768	KFM111Y	772	KFM115Y	776	MTU124Y	780	A138SMA
764	GFM105X	769	KFM112Y	773	MTU120Y	777	MTU125Y	781	A143SMA
765	GFM106X	770	KFM113Y	774	MTU122Y	778	A136SMA	782	A144SMA
766	GFM108X	771	KFM114Y	775	MTU123Y	779	A137SMA	783	A145SMA
767	GFM109X								

DOG784-799   Leyland Olympian ONLXB/1R   Eastern Coach Works   H45/32F   1984-85 Ex Crosville, 1990

784	A146UDM	788	A159UDM	791	A162VDM	794	A165VDM	797	A168VFM
785	A156UDM	789	A160UDM	792	A163VDM	795	A166VFM	798	A169VFM
786	A157UDM	790	A161VDM	793	A164VDM	796	A167VFM	799	A170VFM
787	A158UDM								

SAD801-809   DAF SB220LC550   Optare Delta   DP48F   1990

| 801 | H801GRE | 803 | H803GRE | 805 | H805GRE | 807 | H807GRE | 809 | H809GRE |
| 802 | H802GRE | 804 | H804GRE | 806 | H806GRE | 808 | H808GRE | | |

SLC845	F361YTJ	Leyland Lynx LX112L10ZR1R	Leyland Lynx	B51F	1988	Ex Topp-Line, Wavertree, 1994
SLC846	F362YTJ	Leyland Lynx LX112L10ZR1R	Leyland Lynx	B51F	1988	Ex Topp-Line, Wavertree, 1994
SLC847	F363YTJ	Leyland Lynx LX112L10ZR1R	Leyland Lynx	B51F	1988	Ex Topp-Line, Wavertree, 1994
SLC848	F364YTJ	Leyland Lynx LX112L10ZR1R	Leyland Lynx	B51F	1988	Ex Topp-Line, Wavertree, 1994
SLC849	F608WBV	Leyland Lynx LX112L10ZR1S	Leyland Lynx	B51F	1988	Ex Westbus, Ashford, 1993
SLC850	G136YRY	Leyland Lynx LX112L10ZR1R	Leyland Lynx	B51F	1990	Ex Westbus, Ashford, 1993

The North & West Midlands Bus Handbook

SLC851-861
Leyland Lynx LX2R11C15Z4S　　Leyland Lynx　　　　　　　　DP48F　　1990

851	H851GRE	854	H854GRE	856	H856GRE	858	H858GRE	860	H860GRE
852	H852GRE	855	H855GRE	857	H857GRE	859	H859GRE	861	H861GRE
853	H853GRE								

| SDC862 | L862HFA | Dennis Lance 11SDA3112 | Northern Counties Paladin | DP47F | 1993 |

SDC863-867
Dennis Lance 11SDA3113　　Plaxton Verde　　　　　　　B49F　　1995

| 863 | N863CEH | 864 | N864CEH | 865 | N865CEH | 866 | N866CEH | 867 | N867CEH |

DOG891	A171VFM	Leyland Olympian ONLXB/1R	Eastern Coach Works	H45/32F	1984	Ex Crosville, 1990
DOG892	B181BLG	Leyland Olympian ONLXB/1R	Eastern Coach Works	H45/32F	1984	Ex Crosville, 1990
DOG893	B182BLG	Leyland Olympian ONLXB/1R	Eastern Coach Works	H45/32F	1984	Ex Crosville, 1990
DOG894	B188BLG	Leyland Olympian ONLXB/1R	Eastern Coach Works	H45/32F	1985	Ex Crosville, 1990
DOG895	B195BLG	Leyland Olympian ONLXB/1R	Eastern Coach Works	H45/32F	1985	Ex Crosville, 1990
DOG896	B199DTU	Leyland Olympian ONLXB/1R	Eastern Coach Works	H45/32F	1985	Ex Crosville, 1990
DOG897	B200DTU	Leyland Olympian ONLXB/1R	Eastern Coach Works	DPH42/32F	1985	Ex Crosville, 1990
DOG898	B201DTU	Leyland Olympian ONLXB/1R	Eastern Coach Works	DPH42/32F	1985	Ex Crosville, 1990
DOG899	B202DTU	Leyland Olympian ONLXB/1R	Eastern Coach Works	DPH42/32F	1985	Ex Crosville, 1990
D900	WVT900S	Foden/NC 6LXB	Northern Counties	H43/31F	1978	

IDC901-920
Dennis Dart 9SDL3011　　Plaxton Pointer　　　　　　　DP35F　　1991-92

901	J901SEH	905	J905SEH	909	J909SEH	913	J913SEH	917	J917SEH
902	J902SEH	906	J906SEH	910	J910SEH	914	J914SEH	918	J918SEH
903	J903SEH	907	J907SEH	911	J911SEH	915	J915SEH	919	K919XRF
904	J904SEH	908	J908SEH	912	J912SEH	916	J916SEH	920	K920XRF

IDC921-929
Dennis Dart 9SDL3016　　Plaxton Pointer　　　　　　　DP35F　　1992

| 921 | K921XRF | 923 | K923XRF | 925 | K925XRF | 927 | K927XRF | 929 | K929XRF |
| 922 | K922XRF | 924 | K924XRF | 926 | K926XRF | 928 | K928XRF | | |

IDC930-934
Dennis Dart 9SDL3034　　Plaxton Pointer　　　　　　　DP35F　　1993

| 930 | L930HFA | 931 | L931HFA | 932 | L932HFA | 933 | L933HFA | 934 | L934HFA |

| IDC935 | L935HFA | Dennis Dart 9.8SDL3025 | Marshall C36 | DP36F | 1993 |
| IDC936 | L936HFA | Dennis Dart 9.8SDL3025 | Marshall C36 | DP36F | 1993 |

IDC937-942
Dennis Dart 9SDL3034　　Plaxton Pointer　　　　　　　DP35F　　1994

| 937 | L937LRF | 939 | L939LRF | 940 | L940LRF | 941 | L941LRF | 942 | L942LRF |
| 938 | L938LRF | | | | | | | | |

IDC943-952
Dennis Dart 9SDL3040　　Marshall C36　　　　　　　　DP35F　　1994

| 943 | M943SRE | 945 | M945SRE | 947 | M947SRE | 949 | M949SRE | 952 | M952SRE |
| 944 | M944SRE | 946 | M946SRE | 948 | M948SRE | 951 | M951SRE | | |

IDC953-972
Dennis Dart 9.8SDL3054　　Plaxton Pointer　　　　　　B36F　　1995

953	M953XVT	957	M957XVT	961	M961XVT	965	M965XVT	969	M969XVT
954	M954XVT	958	M958XVT	962	M962XVT	966	M966XVT	970	M970XVT
955	M955XVT	959	M959XVT	963	M963XVT	967	M967XVT	971	M971XVT
956	M956XVT	960	M960XVT	964	M964XVT	968	M968XVT	972	M972XVT

Previous Registrations:
| FXI8653 | B116NBF | | XRF1X | | B115NBF | | XRF2X | | YRE472Y |

Livery: Red and yellow (Crosville, PMT, Red Rider and Pennine); brown and cream (Turners) 430/1, 550/1, 740/2, 928.

The latest arrivals with PMT are Dennis products bodied by the Henly group members, Plaxton. *Opposite top* is Dennis Lance SDC867, N867CEH in a special livery for the Crewe-Hanley service. This bus has a Plaxton Verde body. *Cliff Beeton*

***Opposite, bottom*: Dennis Dart IDC956, M956XVT carries a Plaxton Pointer body and was used to help determine the fleetname styles for FirstBus names and symbols. Seen on the frequent service 24 the vehicle displays a 'round f' with the FirstBus standard font.** *Cliff Beeton*

Alsager Kidsgrove Tunstall
HANLEY
320
EVERY 30 MINUTES
SDC 867
DENNIS
PLAXTON
N867 CEH

Hanley Stoke Longton
MEIR SQUARE
24
PMT
The Potteries Connection
IDC956
PMT
PMT
DENNIS
M956 XVT
PLAXTON

PROCTERS

F Procter & Son Ltd, Dewsbury Road, Fenton, Stoke-on-Trent ST4 2HS

AFA729S	AEC Reliance 6U3ZR	Duple Dominant II	C57F	1978	
YUT326Y	Bristol LH6L	Plaxton Supreme III	C45F	1978	Ex Jalna, Church Gresley, 1981
HRE128V	Leyland Leopard PSU3E/4R	Plaxton Supreme IV Express	C53F	1979	
HRE129V	Leyland Leopard PSU3E/4R	Plaxton Supreme IV Express	C53F	1979	
HIL7624	Leyland Leopard PSU3E/4R	Plaxton Supreme IV Express	C53F	1979	Ex Antler, Rugeley, 1980
WCK128V	Leyland Leopard PSU3E/4R	Duple Dominant II Express	C53F	1979	Ex Cumberland, 1988
HIL7621	Leyland Tiger TRCTL11/3R	Duple Dominant IV	C57F	1982	
HIL7622	Leyland Tiger TRCTL11/3R	Duple Dominant IV	C57F	1982	
HIL7623	Leyland Leopard PSU3E/4R	Plaxton Supreme VI Express	C53F	1982	
HIL2379	Leyland Tiger TRCTL11/3R	Duple Dominant IV	C57F	1982	
EVT690Y	Ford Transit 190	Deansgate	M12	1983	
HIL2376	Leyland Royal Tiger RTC	Leyland Doyen	C49FT	1986	
HIL2377	DAF SB2300DHS585	Duple 340	C57F	1986	Ex Smiths, Alcester, 1987
HIL2378	Leyland Tiger TRCTL11/3RZ	Duple 340	C57F	1986	
HIL2375	DAF SB2300DHS585	Duple 340	C53F	1987	
HIL7620	Scania K112CRB	Van Hool Alizée	C49FT	1987	Ex Stanley Gath, Dewsbury, 1992
HIL7615	Dennis Javelin 12SDA1907	Duple 320	C57F	1988	
HIL7616	Dennis Javelin 12SDA1907	Duple 320	C57F	1988	
HIL7613	Dennis Javelin 12SDA1907	Duple 320	C55F	1989	
HIL7614	Dennis Javelin 12SDA1907	Duple 320	C57F	1989	
HIL7386	DAF MB230LT615	Van Hool Alizée	C53F	1990	Ex Smiths, Alcester, 1992

Previous Registrations:

HIL2375	D294XCX	HIL7386	G977KJX	HIL7620	E58VHL
HIL2376	C269XRF	HIL7613	F702TBF	HIL7621	WFA210X
HIL2377	C780MVH	HIL7614	F703TBF	HIL7622	WFA209X
HIL2378	C962YBF	HIL7615	E846LRF	HIL7623	WVT107X
HIL2379	ARE508Y	HIL7616	E847LRF	HIL7624	JRE355V

Livery: Blue and cream

REST & RIDE

Rest & Ride (Birmingham) Ltd, 1 Rolfe Street, Smethwick, Birmingham B66 2AA

XIB3106	Leyland National 1151/1R/0401		B52F	1973	Ex Walsall Travel, 1994
PCN423M	Leyland National 1151/1R		B52F	1974	Ex MBC1, Rowley Regis, 1995
XPD230N	Leyland National 10351/1R		B39F	1974	Ex Transol, Birmingham, 1995
SEL236N	Leyland National 11351/1R		B49F	1974	Ex Walsall Travel, 1995
HPF298N	Leyland National 10351/1R/SC		DP39F	1975	Ex Walsall Travel, 1995
NFW966P	Leyland National 11351/1R		B49F	1976	Ex Walsall Travel, 1995
KNV505P	Leyland National 11351/1R		B48F	1976	Ex MBC1, Rowley Regis, 1995
PCD75R	Leyland National 11351A/1R		B49F	1976	Ex MTL, 1995
LIW1324	Leyland Leopard PSU3C/2R	Willowbrook	B51F	1976	Ex MBC1, Rowley Regis, 1995
UHG748R	Leyland National 11351A/1R		B49F	1976	Ex MTL, 1995
UTU981R	Leyland National 11351A/1R		B49F	1977	Ex MTL, 1995
VPT945R	Leyland National 11351A/1R		B49F	1977	Ex MBC1, Rowley Regis, 1995
VKE567R	Leyland National 11351A/1R		B49F	1977	Ex MTL, 1995
YEV310S	Leyland National 11351A/1R		B49F	1978	Ex MTL, 1995
XIB3104	Leyland National 11351A/1R		B49F	1978	Ex Walsall Travel, 1995
AAK106T	Leyland National 10351B/1R		B44F	1979	Ex MBC1, Rowley Regis, 1995
SJI5615	Aüwaerter Neoplan N122/3	Aüwaerter Skyliner	CH52/18CT	1981	Ex MBC1, Rowley Regis, 1995
OCN423X	Volvo B10M-61	Duple Goldliner IV	C49FT	1982	Ex Taj, Walsall, 1995
XIB3102	Volvo B10M-61	Jonckheere Jubilee P90	CH49/9FT	1983	Ex MBC1, Rowley Regis, 1995
F38CWY	Mercedes-Benz 811D	Optare StarRider	B26F	1989	Ex Cowie Leaside, 1996

Previous Registrations:

LIW1324	NTX361R	XIB3102	A125SNH, RDU4, A877DUY, SJI5616
LIW1324	NTX361R	XIB3104	CUP669S
SJI5615	STT602X, 4040SC, LES667X, OIA6839	XIB3105	WNO561L

Opposite: **Procters' Leyland Leopard HIL7623 is seen on the operators Hanley to Leek service. The vehicle features a Plaxton Supreme VI body style with shallower windows and Express specification, a combination found on only some 45 bodies.** *Cliff Beeton*

Below: **One of the earlier Leyland Tigers was HIL7621 with Duple Dominant IV bodywork new to Procters in 1982.** *Cliff Beeton*

SANDWELL TRAVEL

Salelink Ltd, Unit 45A, Siddons Estate, Howard Street, West Bromwich,
West Midlands   B70 0TB

TCH275L	Bristol RELH6L	Eastern Coach Works	DP30FL	1973	Ex Coaching at Wheels, 1994
OAF40M	Bedford YRQ	Duple Dominant Express	C45F	1973	Ex Hambly, Pelynt, 1992
TDF224R	Bedford YRT	Plaxton Supreme III	C53F	1977	Ex Dudley Coachways, 1992
SEL530X	Ford R1014	Wadham Stringer Vanguard	B33F	1981	Ex Dorset Health Authority, 1992
BHK565X	Ford R1014	Wadham Stringer Vanguard	B35F	1982	Ex King Alfred School, 1995
JNO52Y	Ford R1115	Wadham Stringer Vanguard	B42F	1983	Ex Roadmark, Storrington, 1993
TDC829X	Mercedes-Benz L307D	Devon Conversions	M12	1982	Ex Alderson-Heil, Wolsingham, 1993
SDW236Y	Dennis Lancet SD512	Wadham Stringer Vanguard	DP33F	1983	Ex Redby, Sunderland, 1993
B841WYH	Leyland Cub CU435	Wadham Stringer Vanguard	B32F	1984	Ex LB Islington, 1993
C805KBT	Leyland Cub CU435	Optare	B33F	1986	Ex Boomerang Bus, Tewkesbury, 1994
D233MKK	Renault-Dodge S56	East Lancashire	B25F	1986	Ex Boro'line, 1991
D234MKK	Renault-Dodge S56	East Lancashire	B25F	1986	Ex Boro'line, 1991
D235MKK	Renault-Dodge S56	East Lancashire	B25F	1986	Ex Boro'line, 1991
D236MKK	Renault-Dodge S56	East Lancashire	B25F	1986	Ex Boro'line, 1991
D735JUB	Freight Rover Sherpa	Dormobile	B20F	1987	Ex Clearway, Catshill, 1992

Livery: Red and grey

West Bromwich bus station is the setting for this view of Sandwell Travel's SDW236Y, a Dennis Lancet with Wadham Stringer Vanguard bodywork. Sandwell operate a similar bodied Leyland Cub, though that is fitted with bus seating. *Richard Godfrey*

SERVEVERSE

Serveverse Ltd, 11 Greenlee, Stoneydelph, Tamworth, Staffordshire

Depot : Mile Oak Business Park, Tamworth

	WNO559L	Leyland National 1151/1R/0401		B52F	1973	Ex LB Wandsworth, 1995
	JOX496P	Leyland National 11351/1R		B49F	1976	Ex Midland Red West, 1993
w	OOX820R	Leyland National 11351A/1R		DP45F	1977	Ex C & H, Fleetwood, 1992
	YCD83T	Leyland National 11351A/2R		B44D	1978	Ex Thames Transit, 1992
	YCD84T	Leyland National 11351A/2R		B44D	1978	Ex Thames Transit, 1992
	AKU165T	Leyland National 10351B/1R		B44F	1979	Ex Thomas of Barry, 1994
	EPD532V	Leyland National 10351B/1R		B44F	1979	Ex Sovereign, 1993
	A573BRD	Dodge S56	Rootes	B22F	1984	Ex ?, 1995
	D525NDA	Freight Rover Sherpa	Carlyle	B19F	1986	Ex Staffordian, Stafford, 1993
	D102AFV	Renault-Dodge S56	Northern Counties	B22F	1987	Ex City Buslines, Birmingham, 1995
	D912NBA	Renault-Dodge S56	Northern Counties	B22F	1987	Ex City Buslines, Birmingham, 1995
	L321BNX	Mercedes-Benz 811D	Plaxton Beaver	B33F	1993	
	M46POL	Mercedes-Benz 811D	Plaxton Beaver	B31F	1995	

Livery: Light green and white

By early May 1995 Serveverse were operating A573BRD on service from their base in Tamworth, where it may still be found. This Rootes-bodied Dodge S66 lacks the destination aperture and this is overcome by boards placed in the windscreen. *Richard Godfrey*

SHROPSHIRE EDUCATION

Shropshire County Council, Shirehall, Abbey Foregate, Shrewsbury SY2 6ND

ETA978Y	Ford R1114	Duple Dominant IV	C53F	1983	Ex Devonways, Kingskerswell, 1983
A653ANT	Ford R1114	Duple Dominant IV	C53F	1983	
B341BBV	Ford R1115	Plaxton Paramount 3200 II	C53F	1985	Ex Stott, Oldham, 1991
H483ONT	Peugeot-Talbot Freeway	Talbot	B16F	1990	
H484ONT	Peugeot-Talbot Pullman	Talbot	B16F	1990	
H485ONT	Peugeot-Talbot Pullman	Talbot	B16F	1990	
H157EFK	Peugeot-Talbot Pullman	Talbot	B16F	1991	
H351FDU	Peugeot-Talbot Pullman	Talbot	B20F	1990	
J3KCB	Mercedes-Benz 609D	Autobus Classique	B24F	1992	
J915HGD	Peugeot-Talbot Pullman	Talbot	B20F	1991	Ex Aryll Coaches, Greenock
L767LAW	Mercedes-Benz 609D	Autobus Classique	B24F	1994	

Livery: White

Shropshire Education is now a commercial business that has also tendered for other service. School feeders still provide most of the services, though special-needs work is undertaken with minibuses. Shown leaving the Charlton School in the Dothill district of Telford is ETA978Y, a Ford R1115 with Duple Dominant bodywork. *Bill Potter*

STEVENSONS

Stevensons of Uttoxeter Ltd, The Garage, Spath, Uttoxeter, Staffordshire   ST14 5AE

Depots : Hot Lane, Burslem; Wetmore Road, Burton-on-Trent; Litchfield; Sunderland Street, Macclesfield; Midland Road, Swadlincote; Ryder Close, Cadley Hill, Swadlincote; The Garage, Spath.

1	PCW946	Volvo B10M-61	Plaxton Paramount 3500	C49FT	1984	Ex Bagnall, Swadlincote, 1989
2	82HBC	DAF MB200DKFL600	Plaxton Paramount 3200	C53F	1983	Ex Viking, Woodville, 1987
3	565LON	Volvo B10M-61	Plaxton Paramount 3200 II	C57F	1985	Ex Blue Bus Services, 1995
4	614WEH	Volvo B58-61	Plaxton P 3200 II (1986)	C53F	1976	Ex Coliseum, Southampton, 1985
5	XAF759	Volvo B10M-61	Plaxton Paramount 3500	C53F	1984	Ex Blue Bus Services, 1995
6	852YYC	Volvo B10M-61	Plaxton Paramount 3500 II	C53F	1985	Ex Blue Bus Services, 1995
7	422AKN	Leyland Leopard PSU3E/4R	Plaxton Supreme III	C53F	1978	
8	AAX562A	Leyland Leopard PSU3E/4R	Plaxton Supreme III Express	C51F	1977	Ex Edinburgh Transport, 1994
9	G417WFP	Bova FHD12.290	Bova Futura	C36FT	1990	Ex Boyden, Castle Donington, 1991
10	803HOM	Volvo B10M-61	Plaxton Paramount 3200 III	C53F	1987	Ex Blue Bus Services, 1995
11	OGL518	Volvo B10M-61	Plaxton Paramount 3200 II	C53F	1985	Ex Blue Bus Services, 1995
12	AAL303A	Leyland Leopard PSU5D/4R (TL11)	Plaxton P3200 III (1987)	C53F	1980	Ex Rhondda, 1992
13	AAL404A	Leyland Leopard PSU5D/4R (TL11)	Plaxton P3200 III (1987)	C53F	1980	Ex Rhondda, 1992
14	LUY742	Volvo B10M-61	Plaxton Paramount 3500 III	C49FT	1987	Ex Sealandair, West Bromwich, 1991
15	VOI6874	Volvo B10M-61	Plaxton Paramount 3500	C53F	1983	Ex Bagnall, Swadlincote, 1989
16	488BDN	Leyland Leopard PSU3E/4R	Plaxton Supreme III Express	C49F	1977	Ex Rhondda, 1993
17	WYR562	Leyland Leopard PSU3E/4R	Plaxton Supreme III Express	C53F	1978	Ex Greater Manchester PTE, 1984
18	479BOC	Leyland Leopard PSU3B/4R (TL11)	Duple 320 (1987)	C53F	1973	Ex Blue Bus, Rugeley, 1985
19	468KPX	Volvo B10M-61	Van Hool Alizée	C44FL	1982	Ex Cumberland, 1992
20	784RBF	Volvo B10M-61	Jonckheere Jubilee P50	C53F	1987	Ex Telling-Golden Miller, 1993
21	G21YVT	Volvo B10M-60	Van Hool Alizée	C53F	1989	
22	G122DRF	Volvo B10M-60	Van Hool Alizée	C49F	1990	
23	HIL3652	Volvo B10M-61	Duple 340	C55F	1987	Ex Crosville Wales, 1995
24	124YTW	Volvo B58-61	Plaxton Supreme IV	C53F	1980	Ex G M Buses, 1986
25	G25YVT	Volvo B10M-60	Plaxton Paramount 3200 III	C53F	1989	
26	XOR841	Volvo B10M-61	Van Hool Alizée	C53F	1983	Ex Sealandair, West Bromwich, 1991
27	TOU962	Volvo B10M-61	Van Hool Alizée	C53F	1983	Ex Sealandair, West Bromwich, 1991
28	FAZ3195	Leyland Tiger TRCTL11/2RH	Plaxton Paramount 3200 II	C49F	1985	Ex Crosville Wales, 1995
29	FAZ5279	Leyland Tiger TRCTL11/3R	Plaxton Paramount 3200 E	C53F	1984	Ex Crosville Wales, 1995
30	C252SPC	Leyland Tiger TRCTL11/3R	Duple 340	C53F	1986	Ex Crosville Wales, 1996
31	J31SFA	Leyland Swift ST2R44C97A4	Wright Handy-bus	B39F	1992	
32	J32SFA	Leyland Swift ST2R44C97A4	Wright Handy-bus	B39F	1992	
33	H313WUA	Leyland Swift ST2R44C97A4	Reeve Burgess Harrier	DP39F	1991	Ex Pennine, Gargrave, 1992
34	J34SRF	Leyland Swift ST2R44C97A4	Wright Handy-bus	B39F	1992	
35	H314WUA	Leyland Swift ST2R44C97A4	Reeve Burgess Harrier	DP39F	1991	Ex Pennine, Gargrave, 1992
36	J36SRF	Leyland Swift ST2R44C97A4	Wright Handy-bus	B39F	1992	

Stevensons 18, 479BOC is seen in Derby bus station with dedicated livery for the half-hourly express between there and Burton. The Duple 320 body was built in 1987 onto a Leyland Leopard chassis then fourteen years old.
David Stanier

85

The recent transfer of North Western's express coaches to Stevensons is illustrated in this view of Duple Laser 45, B149ALG caught by the camera at Burton while operating school duties from its base at Uttoxeter. *Tony Wilson*

37	G616WGS	Leyland Swift LBM6T/2RA	Reeve Burgess Harrier	B39F	1989	Ex Chambers, Stevenage, 1992
38	F907PFH	Leyland Swift LBM6T/2RA	G C Smith Whippet	B36F	1988	Ex Gloucestershire CC, 1993
39	G727RGA	Leyland Swift LBM6T/2RA	Reeve Burgess Harrier	B39F	1990	Ex Kelvin Central, 1993
40	YSU953	Leyland Tiger TRCTL11/3RH	Plaxton Paramount 3200 E	C53F	1983	Ex Frontline, 1996
41	YSU954	Leyland Tiger TRCTL11/3RH	Plaxton Paramount 3200 E	C53F	1983	Ex Frontline, 1996
42	OKY822X	Leyland Leopard PSU5C/4R	Plaxton Supreme VI	C57F	1982	Ex Frontline, 1996
43	B147ALG	Leyland Tiger TRCTL11/2RH	Duple Laser 2	C49F	1984	Ex North Western, 1995
44	B148ALG	Leyland Tiger TRCTL11/2RH	Duple Laser 2	C49F	1984	Ex North Western, 1995
45	B149ALG	Leyland Tiger TRCTL11/2RH	Duple Laser 2	C49F	1984	Ex North Western, 1995
46	B150ALG	Leyland Tiger TRCTL11/2RH	Duple Laser 2	C49F	1984	Ex North Western, 1995
47	A39SMA	Leyland Tiger TRCTL11/2RH	Duple Laser	C49F	1983	Ex North Western, 1995
48	BOK68V	MCW Metrobus DR102/12	MCW	H43/30F	1980	Ex West Midlands Travel, 1990
49	GOG223W	MCW Metrobus DR102/18	MCW	H43/30F	1981	Ex West Midlands Travel, 1990
50	GOG272W	MCW Metrobus DR102/18	MCW	H43/30F	1981	Ex West Midlands Travel, 1990

51-56 — MCW Metrobus DR102/22 — MCW — H43/30F — 1981 — Ex West Midlands Travel, 1990

51	KJW296W	53	KJW305W	54	KJW306W	55	KJW310W	56	KJW322W
52	KJW301W							

57	GOG168W	MCW Metrobus DR102/18	MCW	H43/30F	1981	Ex North Western, 1995
58	JHE137W	MCW Metrobus DR104/6	MCW	H46/31F	1981	Ex South Yorkshire's Transport, 1990
59	LOA429X	MCW Metrobus DR102/22	MCW	H43/30F	1982	Ex North Western, 1995
60	D401MHS	Leyland Lynx LX5636LXCTFR1	Leyland Lynx	B47F	1986	Ex Kelvin Central, 1991
61	F61PRE	Leyland Lynx LX112L10ZR1R	Leyland Lynx	B48F	1989	
62	B145ALG	Leyland Tiger TRCTL11/2RH	Duple Laser 2	C49F	1984	Ex North Western, 1995
63	MFR126P	Leyland Leopard PSU4C/2R	Alexander AYS	B45F	1976	Ex Edinburgh Transport, 1994
64	MFR125P	Leyland Leopard PSU4C/2R	Alexander AYS	B45F	1976	Ex Edinburgh Transport, 1994
65	MFR41P	Leyland Leopard PSU4C/2R	Alexander AY	B45F	1976	Ex Edinburgh Transport, 1994

Opposite: The Stevensons operation is now receiving a similar livery to sister company Midland Red North who operate under the Midland name. The enlarged management have two divisions, Midland's Tamworth operations are now part of the eastern division that also includes those of Stevensons, though the fleets have yet to be brought into one. Comparison of the new and former liveries is seen here with Metrobus 81, F181YDA and 306, J556GTP, a Dennis Dart with Wadham Stringer Portsdown bodywork. *Tony Wilson*

5
WINSHILL
Stevensons
81
STEVENSONS
F181 YDA

14
ANGLESEY ROAD
Stevensons
306
STEVENSONS
J556 GTP

Mercedes-Benz have made several attempts to enter the British bus single-deck market but the high cost of the import and a strong home market for British-built Volvo, Optare and Dennis has restricted sales. In recent years the Mercedes-Benz 0405 has become available in chassis form, first with Alexander bodywork and latterly Wright and Optare have built bodies. Seen as Stevensons 100, L100SBS is this example with Wright Cityranger bodywork that includes the Mercedes-Benz front. *Tony Wilson*

66	H166MFA	Leyland Swift ST2R44C97A4	Wadham Stringer Vanguard II B39F		1991	
67	F956XCK	Leyland Swift LBM6N/2RAO	Wadham Stringer Vanguard II B39F		1989	Ex Jim Stones, Glazebury, 1991
68	G98VMM	Leyland Swift LBM6T/2RA	Wadham Stringer Vanguard II B39F		1989	Ex Green, Kirkintilloch, 1991
69	J169REH	Leyland Swift ST2R44C97A4	Wadham Stringer Vanguard II B39F		1991	
70	KJW318W	MCW Metrobus DR102/22	MCW	H43/30F	1981	Ex West Midlands Travel, 1990
71	KJW320W	MCW Metrobus DR102/22	MCW	H43/30F	1981	Ex West Midlands Travel, 1990
72	LOA388X	MCW Metrobus DR102/22	MCW	H43/30F	1981	Ex North Western, 1995
73	UWW512X	MCW Metrobus DR101/15	Alexander RH	H43/32F	1982	Ex Yorkshire Rider, 1987
74	UWW513X	MCW Metrobus DR101/15	Alexander RH	H43/32F	1982	Ex Yorkshire Rider, 1987
75	UWW515X	MCW Metrobus DR101/15	Alexander RH	H43/32F	1982	Ex Yorkshire Rider, 1987
76	UWW517X	MCW Metrobus DR101/15	Alexander RH	H43/32F	1982	Ex Yorkshire Rider, 1987
77	JWF490W	MCW Metrobus DR102/13	MCW	H46/30F	1980	Ex South Yorkshire's Transport, 1988
78	GBU7V	MCW Metrobus DR101/6	MCW	H43/30F	1979	Ex GM Buses, 1986
79	BSN878V	MCW Metrobus DR102/5	MCW	H45/30F	1979	Ex Enterprise & Silver Dawn, 1988
80	TOJ592S	MCW Metrobus DR101/2	MCW	H43/30F	1977	Ex MCW demonstrator, 1989
81	F181YDA	MCW Metrobus DR132/12	MCW	H43/30F	1988	Ex MCW demonstrator, 1989
82	BOK72V	MCW Metrobus DR102/12	MCW	H43/30F	1980	Ex West Midlands Travel, 1989
83	BOK75V	MCW Metrobus DR102/12	MCW	H43/30F	1980	Ex West Midlands Travel, 1989
85	EEH902Y	Leyland Olympian ONLXB/1R	Eastern Coach Works	H45/32F	1983	Ex Midland, 1994
86	EEH904Y	Leyland Olympian ONLXB/1R	Eastern Coach Works	H45/32F	1983	Ex Midland, 1994
87	AHW206V	Bristol VRT/SL3/6LXB	Eastern Coach Works	H43/30F	1980	Ex Frontline, 1996
88	A152UDM	Leyland Olympian ONLXB/1R	Eastern Coach Works	H45/32F	1984	Ex Midland, 1994
89	D676MHS	MCW Metrobus DR102/52	Alexander RL	DPH45/33F	1986	Ex Kelvin Central, 1994
90	D678MHS	MCW Metrobus DR102/52	Alexander RL	DPH45/33F	1986	Ex Kelvin Central, 1994
91	D680MHS	MCW Metrobus DR102/52	Alexander RL	DPH45/33F	1986	Ex Kelvin Central, 1994
92	D682MHS	MCW Metrobus DR102/52	Alexander RL	H45/33F	1986	Ex Kelvin Central, 1994
93	D683MHS	MCW Metrobus DR102/52	Alexander RL	DPH45/33F	1986	Ex Kelvin Central, 1994
94	L94HRF	DAF DB250RS200505	Optare Spectra	H48/29F	1993	
95	L95HRF	DAF DB250RS200505	Optare Spectra	H48/29F	1993	
96	F96PRE	Leyland Olympian ONCL10/1RZ	Alexander RL	H47/32F	1988	
97	F97PRE	Leyland Olympian ONCL10/1RZ	Alexander RL	H47/32F	1988	
98	AHW207V	Bristol VRT/SL3/6LXB	Eastern Coach Works	H43/30F	1980	Ex Frontline, 1996
99	Q246FVT	Leyland Olympian B45-6LXB	Eastern Coach Works	H43/30F	1979	Ex Leyland development, 1983

Several Willowbrook Warrior re-bodies are operated from the Stevensons fleet, two having been acquired with the Frontline business. Seen in Burton-on-Trent while heading for Swadlincote is 125, GDZ795 which migrated from Allanders' Loch Lomond Coaches operation. *Tony Wilson*

100	L100SBS	Mercedes-Benz 0405	Wright Cityranger	B51F	1993	
101	PUA315W	Leyland Atlantean AN68B/1R	Roe	H43/32F	1981	Ex Frontlinr, 1996
102	L102MEH	MAN 11.190	Optare Vecta	B42F	1994	
103	K140RYS	MAN 11.190	Optare Vecta	B37F	1993	Ex Express Travel, Perth, 1994
104	K141RYS	MAN 11.190	Optare Vecta	B40F	1993	Ex Express Travel, Perth, 1994
105	E829AWA	Leyland TRBTL11/2RP	Plaxton Derwent II	B54F	1988	Ex Liverline, 1993
106	H408YMA	Leyland Lynx LX2R11C15Z4R	Leyland Lynx	B51F	1990	Ex The Wright Company, Wrexham, 1994
107	F170DET	Scania K93CRB	Plaxton Derwent II	B57F	1989	Ex Capital Citybus, 1993
108	F258GWJ	Leyland Lynx LX112L10ZR1R	Leyland Lynx	B51F	1989	Ex The Wright Company, Wrexham, 1993
109	G109YRE	Scania K93CRB	Alexander PS	B51F	1989	
110	F110SRF	Scania K93CRB	Alexander PS	B51F	1989	
111	VAJ785S	Leyland Leopard PSU3E/4R	Willowbrook Warrior (1990)	B48F	1977	Ex South Lancs Transport, 1994
112	E72KBF	Leyland Lynx LX112L10ZR1	Leyland Lynx	B51F	1988	
113	YSF85S	Leyland Leopard PSU3D/4R	Alexander AYS	B53F	1977	Ex Fife Scottish, 1992
114	OVT344P	Leyland Leopard PSU3C/4R	Plaxton Elite III	C53F	1977	Ex Selwyn, Runcorn, 1983
115	YYJ955	Leyland Leopard PSU3E/4R	Willowbrook Warrior(1992)	B53F	1978	Ex Frontline, 1996
117	904AXY	Leyland Leopard PSU3E/4R	Willowbrook Warrior(1992)	B48F	1978	Ex Frontline, 1996
118	A41SMA	Leyland Tiger TRCTL11/2R	Duple Laser	C49F	1983	Ex North Western, 1995
119	UOI772	Leyland Leopard PSU3C/4R	Willowbrook Warrior(1989)	B53F	1978	Ex Loch Lomond Coaches, 1994
120	GMS295R	Leyland Leopard PSU3C/4R	Alexander AYS	B53F	1978	Ex Henley's, Abertillery, 1991
121	JUM531V	Leyland Leopard PSU3E/4R	Plaxton Supreme III Express	C53F	1979	Ex Frontline, 1996
122	KUB671V	Leyland Leopard PSU3E/4R	Plaxton Supreme III Express	C49F	1980	Ex Frontline, 1996
123	BJT322T	Leyland Leopard PSU3E/4R	Plaxton Supreme III Express	C49F	1979	Ex Frontline, 1996
124	MFR18P	Leyland Leopard PSU3C/4R	Alexander AY	B53F	1976	Ex Edinburgh Transport, 1984
125	GDZ795	Leyland Leopard PSU3C/4R	Willowbrook Warrior(1989)	B53F	1975	Ex Loch Lomond Coaches, 1994
126	A195KKF	Leyland Tiger TRCTL11/2R	Duple Laser	DP49F	1983	Ex North Western, 1995
127	F77ERJ	Mercedes-Benz 609D	Reeve Burgess Beaver	B27F	1988	Ex Star Line, Knutsford, 1991
128	D133NUS	Mercedes-Benz L608D	Alexander	B21F	1986	Ex Kelvin Central, 1992
129	D135NUS	Mercedes-Benz L608D	Alexander	B21F	1986	Ex Kelvin Central, 1992
130	D141NUS	Mercedes-Benz L608D	Alexander	B21F	1986	Ex Kelvin Central, 1992

The North & West Midlands Bus Handbook

131	D534FAE	Mercedes-Benz L608D	Dormobile	B20F	1986	Ex Frontline, 1996	
132	D538FAE	Mercedes-Benz L608D	Dormobile	B20F	1986	Ex Frontline, 1996	
134	F822GDT	Mercedes-Benz 811D	Reeve Burgess Beaver	C25F	1989	Ex Gordons, Rotherham, 1993	
135	G807FJX	Mercedes-Benz 811D	PMT Ami	DP33F	1990	Ex Traject, Halifax, 1993	
136	K136ARE	Mercedes-Benz 709D	Wright	B29F	1992		
137	K137ARE	Mercedes-Benz 709D	Wright	B29F	1992		
138	K138BRF	Mercedes-Benz 811D	Dormobile Routemaker	B31F	1993		
139	K139BRF	Mercedes-Benz 811D	Dormobile Routemaker	B31F	1993		
140	K140BFA	Mercedes-Benz 811D	Dormobile Routemaker	B31F	1993		
141	K141BFA	Mercedes-Benz 811D	Dormobile Routemaker	B31F	1993		
142	K142BFA	Mercedes-Benz 811D	Dormobile Routemaker	B31F	1993		
143	J143SRF	Mercedes-Benz 709D	Wright	B29F	1992		
144	IDZ8561	Mercedes-Benz 811D	Wright Nim-bus	B26F	1990	Ex Wright demonstrator, 1992	
147	K947BRE	Mercedes-Benz 811D	Dormobile Routemaker	B29F	1993		
149	G901MNS	Mercedes-Benz 811D	Reeve Burgess Beaver	B33F	1989	Ex Edinburgh Transport, 1994	
150	K150BRF	Mercedes-Benz 709D	Wright	B29F	1992		
159	E564YBU	Mercedes-Benz 709D	Reeve Burgess Beaver	B25F	1988	Ex Star Line, Knutsford, 1990	

164-173

Mercedes-Benz 709D LHE B29F 1990

164	G164YRE	166	G166YRE	168	G168YRE	170	G170YRE	172	G172YRE
165	G165YRE	167	G167YRE	169	G169YRE	171	G171YRE	173	G173YRE

176	H176JVT	Mercedes-Benz 811D	Wright	B29F	1990		
177	H177JVT	Mercedes-Benz 811D	Wright	B29F	1990		
178	C78WRE	Mercedes-Benz L608D	PMT Hanbridge	DP19F	1986		
181	D176LNA	Mercedes-Benz 609D	Made-to-Measure	B27F	1986	Ex Marriott, Clayworth, 1988	
183	G183DRF	Mercedes-Benz 709D	LHE	B29F	1990		
184	G184DRF	Mercedes-Benz 709D	LHE	B29F	1990		
185	F185PRE	Mercedes-Benz 709D	Robin Hood	B29F	1988		
186	F186PRE	Mercedes-Benz 709D	Reeve Burgess Beaver	B25F	1988		
187	F187REH	Mercedes-Benz 609D	Whittaker Europa	B20F	1988		
188	F188REH	Mercedes-Benz 609D	PMT	B21F	1988		
189	F189RRF	Mercedes-Benz 709D	Robin Hood	B29F	1988		
190	F190RRF	Mercedes-Benz 709D	Robin Hood	B29F	1988		
191	F191SRF	Mercedes-Benz 709D	Robin Hood	B29F	1989		
192	F192VFA	Mercedes-Benz 709D	Robin Hood	B29F	1989		
193	F326PPO	Mercedes-Benz 709D	Robin Hood	B29F	1989	Ex Robin Hood demonstrator, 1989	
194	B882HSX	Mercedes-Benz L608D	Stevenson	B21F	1984	Ex Scottish C for Spastics, 1989	
197	H197JVT	Mercedes-Benz 814D	Wright Nim-Bus	B33F	1990		
198	H198JVT	Mercedes-Benz 814D	Wright Nim-Bus	B33F	1990		
199	H199KEH	Mercedes-Benz 814D	Phoenix	DP31F	1990		
200	PSV323	MCW MetroRider MF154/2	MCW	C28F	1990	Ex Northumbria, 1994	

201-207

Mercedes-Benz 814D Wright Nim-Bus B31F* 1991 *201/2 are B33F

201	H201LRF	203	J203REH	205	J205REH	206	J206REH	207	J207REH
202	H202LRF	204	J204REH						

208	J208SRF	Mercedes-Benz 709D	Wright Nim-Bus	B27F	1992		
209	J209SRF	Mercedes-Benz 709D	Wright Nim-Bus	B27F	1992		
222	H880NFS	Mercedes-Benz 709D	PMT Ami	B29F	1991	Ex Gold Circle, Airdrie, 1994	
223	C802SDY	Mercedes-Benz L608D	Alexander	B20F	1986	Ex Rainworth Travel, 1993	
224	C823SDY	Mercedes-Benz L608D	Alexander	B20F	1986	Ex East Midland, 1993	
225	C822SDY	Mercedes-Benz L608D	Alexander	B20F	1986	Ex East Midland, 1993	
226	L226JFA	Mercedes-Benz 709D	Dormobile Routemaker	B29F	1993		
229	L229JFA	Mercedes-Benz 709D	Dormobile Routemaker	B27F	1993		
230	L230JFA	Mercedes-Benz 709D	Dormobile Routemaker	B27F	1993		
232	L232JFA	Mercedes-Benz 709D	Dormobile Routemaker	B27F	1993		
241	E478NSC	Mercedes Benz 709D	Alexander Sprint	DP25F	1988	Ex Oakley Buses, 1994	
245	F985EDS	Mercedes Benz 811D	Alexander Sprint	DP31F	1988	Ex Rhondda, 1994	
253	L253NFA	Mercedes-Benz 709D	Wadham Stringer Wessex II	B29F	1994		
254	L254NFA	Mercedes-Benz 709D	Wadham Stringer Wessex II	B29F	1994		
255	L255NFA	Mercedes-Benz 709D	Wadham Stringer Wessex II	B29F	1994		
256	HXI3006	Leyland Lynx LX5636LXCTFR	Alexander N	B49F	1985	Ex Citybus, Belfast, 1992	
257	HXI3007	Leyland Lynx LX5636LXBFR	Alexander N	B49F	1986	Ex Citybus, Belfast, 1992	
258	HXI3008	Leyland Lynx LX5636LXBFR	Alexander N	B49F	1986	Ex Citybus, Belfast, 1992	
259	HXI3009	Leyland Lynx LX5636LXBFR	Alexander N	B49F	1986	Ex Citybus, Belfast, 1992	
260	HXI3010	Leyland Lynx LX563TL11FR	Alexander N	B49F	1986	Ex Citybus, Belfast, 1992	
261	HXI3011	Leyland Lynx LX563TL11FR	Alexander N	B53F	1986	Ex Citybus, Belfast, 1992	
262	HXI3012	Leyland Lynx LX563TL11FR	Alexander N	B53F	1986	Ex Citybus, Belfast, 1992	

266	G785PWL	DAF SB220LC550		Optare Delta	B49F	1989	Ex Edinburgh Transport, 1994
267	F792DWT	DAF SB220LC550		Optare Delta	B49F	1989	Ex Edinburgh Transport, 1994
268	MHJ722V	Leyland National 2 NL116L11/1R			B49F	1980	Ex Frontline, 1996
269	MHJ725V	Leyland National 2 NL116L11/1R			B49F	1980	Ex Frontline, 1996
270	MHJ727V	Leyland National 2 NL116L11/1R			B49F	1980	Ex Frontline, 1996
271	STW18W	Leyland National 2 NL116L11/1R			B49F	1980	Ex Frontline, 1996
272	STW20W	Leyland National 2 NL116L11/1R			B49F	1980	Ex Frontline, 1996
273	NTC625M	Leyland National 1151/1R/0401			B49F	1973	Ex Midland, 1995
300	L300SBS	Dennis Dart 9.8SDL3035		Plaxton Pointer	B40F	1994	
301	L301NFA	Dennis Dart 9.8SDL3035		Plaxton Pointer	B40F	1994	
302	L302NFA	Dennis Dart 9.8SDL3035		Plaxton Pointer	B40F	1994	
303	L303NFA	Dennis Dart 9.8SDL3035		Plaxton Pointer	B40F	1994	
304	L304NFA	Dennis Dart 9.8SDL3035		Plaxton Pointer	B40F	1994	
305	L305NFA	Dennis Dart 9.8SDL3035		Plaxton Pointer	B40F	1994	
306	J556GTP	Dennis Dart 9SDL3002		Wadham Stringer Portsmouth	B35F	1991	Ex Irwell Valley, Boothstown, 1992
307	G141GOL	Dennis Dart 9SDL3002		Duple Dartline	B36F	1990	Ex Star Line, Knutsford, 1992
308	H851NOC	Dennis Dart 9.8SDL3004		Carlyle Dartline	B43F	1991	Ex Thanet Bus, Ramsgate, 1992
309	H192JNF	Dennis Dart 9SDL3002		Wadham Stringer Portsmouth	B35F	1990	Ex Jim Stones, Glazebury, 1993
320	F155DKU	Leyland Swift LBM6T/2RA		Reeve Burgess Harrier	B39F	1989	Ex K-Line, Kirkburton, 1993
321	E990NMK	Leyland Swift LBM6T/2RS		Wadham Stringer Vanguard II	B37F	1988	Ex Armchair, Brentford, 1993
322	E992NMK	Leyland Swift LBM6T/2RS		Wadham Stringer Vanguard II	B37F	1988	Ex Armchair, Brentford, 1993
323	E993NMK	Leyland Swift LBM6T/2RS		Wadham Stringer Vanguard II	B37F	1988	Ex Armchair, Brentford, 1993
324	J162REH	Leyland Swift ST2R44C97A4		Wadham Stringer Vanguard II	B39F	1991	

Previous Registrations:

124YTW	DEN247W	AAX562A	OTD828R	OGL518	B912SPR
422AKN	XRE305S	FAZ3195	B269KPF	OVT344P	LMA61P, YSG339
468KPX	VRR447, UHH575X	FAZ5279	A145EPA	PCW946	A703OWY, HIJ3652
479BOC	AJA360L	GDZ795	LPT903P	PSV323	G298SKP
488BDN	SOA676S	H192JNF	H1JYM	TOU962	MSU573Y
565LON	B549AMH, MSU432	HIJ3652	E472BTN	UOI772	VCA995R
614WEH	LOT777R	HXI3006	From new	VOI6874	YNN29Y
784RBF	D319VVV	HXI3007	From new	WYR562	TWH687T
803HOM	D264MFX	HXI3008	From new	XAF759	B555HAL
82HBC	JGL53, DFP707Y	HXI3009	From new	XOR841	MHS665Y
852YYC	B666KVO	HXI3010	From new	YSU953	A618ATV
904AXY	VRM620S	HXI3011	From new	YSU954	A622ATV
A195KKF	A43SMA, 1205FM	HXI3012	From new	YYJ955	XBF60S
AAL303A	BUH226V	IDZ8561	From new		
AAL404A	BUH222V	LUY742	E562UHS		

Livery: Yellow and black; Victoria travel - Maroon/pink; Viking Coaches - two-tone grey.

Seven Alexander-bodied Leyland Lynx from the Belfast operator, Citybus joined the fleet in 1992 and have retained their index marks from the province. Seen in Uttoxeter is 256, HXI3006, and is particularly interesting in being the first of the Lynx prototype chassis. More information on the Lynx can be found in sister publication, the Leyland Lynx Bus Handbook.
Tony Wilson

TRAVEL DE COURCEY

Mike De Courcey Travel Ltd, Rowley Drive, Coventry   CV3 4FG

JHA246L	Leyland Leopard PSU3B/2R	Marshall	B49F	1973	Ex Excelsior, Telford, 1987
SMU924N	Daimler Fleetline CRL6	Park Royal	H45/32F	1974	Ex Coach Services, Thetford, 1989
MIB9298	Leyland Leopard PSU3C/4R	Willowbrook Warrior(1990)	B53F	1976	Ex Elsey, Gosberton, 1995
KIB6993	Leyland Leopard PSU3D/4R	Willowbrook Crusader (1988)	C50F	1976	Ex Meyric, Magor, 1987
OJD129R	Leyland Fleetline FE30AGR	Park Royal	H45/32F	1976	Ex Graham's, Paisley, 1988
OJD207R	Leyland Fleetline FE30AGR	MCW	H45/32F	1977	Ex Thamesdown, 1994
OJD210R	Leyland Fleetline FE30AGR	MCW	H45/32F	1977	Ex Thamesdown, 1994
OJD215R	Leyland Fleetline FE30AGR	MCW	H45/32F	1977	Ex Thamesdown, 1994
OJD220R	Leyland Fleetline FE30AGR	MCW	H45/32F	1977	Ex Thamesdown, 1995
WFH169S	Leyland Leopard PSU3E/4R	Plaxton Supreme III	C53F	1978	Ex McAndrew, Leamington, 1992
WCJ500T	Volvo B58-56	Plaxton Supreme III Express	C53F	1978	Ex Long, Freshwater, 1995
MJI2369	Leyland Leopard PSU3E/4R	Plaxton Supreme III	C53F	1978	Ex Reliance, Gravesend, 1989
MJI2368	Leyland Leopard PSU3E/4R	Plaxton Supreme IV Express	C53F	1979	Ex Brentwood Coaches, 1987
MJI7863	Leyland Leopard PSU5C/4R	Plaxton Supreme IV	C57F	1979	Ex Wessex, 1986
BVP802V	Leyland Leopard PSU3E/4R	Willowbrook 003	C49F	1980	Ex Treadwell, Barton, 1994
MJI2364	Leyland Leopard PSU3E/4R	Duple Dominant II	C53F	1982	Ex Parnaby, Tolworth, 1987
MJI2370	Leyland Tiger TRCTL11/2R	Plaxton Supreme V Express	C53F	1982	Ex Enterprise, Coventry, 1989
MJI2365	Leyland Tiger TRCTL11/3R	Duple Dominant IV	C55F	1983	Ex Westbus, 1991
MJI2366	Leyland Tiger TRCTL11/3R	Duple Laser	C43FL	1983	Ex Busways, 1993
HHJ378Y	Leyland Tiger TRCTL11/2R	Alexander TE	C53F	1983	Ex Eastern National, 1995
A692CHJ	Leyland Tiger TRCTL11/2R	Alexander TE	C53F	1983	Ex Eastern National, 1995
ESU241	Leyland Tiger TRCTL11/3R	Plaxton Paramount 3500	C49FT	1984	Ex Armchair, Brentford, 1991
A174NAC	Leyland Tiger TRCTL11/3R	Plaxton Paramount 3500	C48FT	1984	Ex Lightfoot, Winsford, 1990
MJI7862	Leyland Tiger TRCTL11/3R	Plaxton Paramount 3200	C53F	1984	Ex Northumbria, 1990
PSU606	Volvo B10M-61	Jonckheere Jubilee	C49FT	1984	Ex Admiral, Welwyn Garden City, 1994
C914AWK	Leyland Tiger TRCTL11/2R	Plaxton Paramount 3500 II	C49FT	1986	Ex Toohey, Nenagh, 1995
86TS330	Leyland Tiger TRCTL11/2R	Plaxton Paramount 3500 II	C49FT	1986	Ex Toohey, Nenagh, 1995
86TN171	Leyland Tiger TRCTL11/3RZ	Caetano Algarve	C51FT	1986	Ex Toohey, Nenagh, 1995
86TN172	Leyland Tiger TRCTL11/3RZ	Caetano Algarve	C51FT	1986	Ex Toohey, Nenagh, 1995
FIL8602	Leyland Tiger TRCTL11/3RZ	Plaxton Paramount 3200 II	C52F	1986	Ex Armchair, Brentford, 1994
MJI2367	Volvo B10M-61	Caetano Algarve	C50FT	1987	Ex Davies, Slough, 1994
MJI7861	Volvo B10M-61	Plaxton Paramount 3500 III	C49FT	1988	Ex Limebourne, Battersea, 1994
MJI4838	Volvo B10M-61	Plaxton Paramount 3500 III	C49FT	1988	Ex Limebourne, Battersea, 1994
88TN1224	Van Hool T815	Van Hool Alizée	C49FT	1988	Ex Toohey, Nenagh, 1995
55MDT	Leyland Tiger TRCTL11/3RZ	Plaxton Paramount 3200 III	C57F	1988	Ex Parfitt's, Rhymney Bridge, 1995
88KK1114	TAZ Dubrava	TAZ D3500	C53F	1988	Ex Toohey, Nenagh, 1995
F811RJF	TAZ Dubrava	TAZ D3500	C49FT	1988	Ex Westbay, Crewe, 1994
G594EKV	Volvo B10M-60	Ikarus Blue Danube	C53F	1991	Ex Toohey, Nenagh, 1995
M288OUR	Iveco 480.10.21	Wadham Stringer Vanguard II	B47F	1994	
M291OUR	Iveco 480.10.21	Wadham Stringer Vanguard II	B47F	1994	

Previous Registrations:
55MDT	E281STG, JEP411, E783XHB	MJI2364	DBH454X
88KK1114	88KK1300	MJI2365	JNM752Y
86TN171	C182RVV	MJI2366	TTN12Y, 552UTE, WCN962Y
86TN172	C183RVV	MJI2367	D702OCY
86TS330	C261GUH	MJI2368	OGR50T
88TN1224	KOU480X, YSV908	MJI2369	WFH181S
A174NAC	82LUP, A828MRW, MJI7861	MJI2370	WBF718X
C914AWK	C260GUH, 86TS329	MJI4838	E304OMG
ESU241	A827PPP	MJI7861	E312OMG
FIL8602	C406DML	MJI7862	A69NPP, WSV573, A674DCN
G594EKV	90D24676	MJI7863	FDF264T
KIB6993	PJF12R	PSU606	B22URU

Livery: White, orange and blue

Travel de Courcey operate several services in Warwickshire, many of which are tendered to that county council. Recently added to the fleet are two Iveco buses, produced in Italy. Bodied by Wadham Stringer, M291OUR is seen in Coventry bus station. In contrast, also in Coventry but working a town service is Leyland Leopard MJI2368 working its way through the centre shopping area. *Bill Potter*

WARRINGTONS

WN & SM Warrington, The Cottage, Ilam, via Ashbourne, DE6 2AZ

SFP829X	Bedford SB5	Duple Dominant	C41F	1981	
A529DNR	Mercedes-Benz L307D	Reeve Burgess	M12	1983	
H153DJU	Dennis Javelin 11SDL1921	Plaxton Paramount 3200 III	C53F	1990	
J733KBC	Dennis Javelin 11SDL1921	Plaxton Paramount 3200 III	C53F	1991	
K2KHW	Leyland DAF 400	Autobus Classique	M16	1992	
K4GWC	Mercedes-Benz 814D	Autobus Classique	C31F	1992	Ex Parker, Nottingham, 1994
L545MRA	Ford Transit VE6	Ford/TCH	M14	1993	

Previous Registrations:
K2KHW   K959EWF

Livery: Red and cream

Warringtons' depot is in the Dove Dale village of Ilam just yards from the Derbyshire border, and while in Staffordshire has an Ashbourne address. The mainstay of the twice-weekly service to Ashbourne is the Bedford SB5, illustrated in the last edition. Seen here in the village is the latest arrival, K4GWC with an Autobus Classique body based on the Mercedes-Benz 814 chassis. *Bill Potter*

W M BUSES

West Midlands Travel Ltd, St Andrews House, 10 St Paul's Square,
Birmingham B3 1QU
Central Coachways (Walsall) Ltd, 67 Oscott Road, Perry Barr, Birmingham
Smiths Coaches (Shennington) Ltd, Miller Street, Aston, Birmingham

Depots : Summer Road, Acocks Green; Miller Street, Aston; Liverpool Street, Birmingham; Wheatley Street, Coventry; Whitmore Street, Hockley; Crossfield Road, Lea Hall; Wellhead Lane, Perry Barr; Ridgeacre Lane, Quinton; Bloxwich Road, Walsall; Washwood Heath Road, Washwood Heath; Oak Lane, West Bromwich; Park Lane, Wolverhampton and Yardley Wood Road, Yardley Wood.

Your Bus:

2	G215HCP	DAF SB220LC550	Optare Delta	B49F	1990	
3	G216HCP	DAF SB220LC550	Optare Delta	B49F	1990	
4	G217HCP	DAF SB220LC550	Optare Delta	B49F	1990	
5	G218HCP	DAF SB220LC550	Optare Delta	B49F	1990	
6	F372KBW	DAF SB220LC550	Optare Delta	DP49F	1989	Ex DAF demonstrator, 1990
7	F335RWK	Leyland Tiger TRBTL11/2RP	Plaxton Derwent II	B54F	1988	
8	F336RWK	Leyland Tiger TRBTL11/2RP	Plaxton Derwent II	B54F	1988	
9	F337RWK	Leyland Tiger TRBTL11/2RP	Plaxton Derwent II	B54F	1988	
10	F338RWK	Leyland Tiger TRBTL11/2RP	Plaxton Derwent II	B54F	1988	
11	E915NAC	Leyland Tiger TRBTL11/2RP	Plaxton Derwent II	B54F	1988	
12	E916NAC	Leyland Tiger TRBTL11/2RP	Plaxton Derwent II	B54F	1988	
13	E917NAC	Leyland Tiger TRBTL11/2RP	Plaxton Derwent II	B54F	1988	
14	E918NAC	Leyland Tiger TRBTL11/2RP	Plaxton Derwent II	B54F	1988	
15	H203TCP	DAF SB220LC550	Ikarus CitiBus	B50F	1991	Ex Hughes DAF, 1992
16	H516YCX	DAF SB220LC550	Ikarus CitiBus	B50F	1991	
17	H517YCX	DAF SB220LC550	Ikarus CitiBus	B50F	1991	
18	J34GCX	DAF SB220LC550	Ikarus CitiBus	B50F	1992	
19	J37GCX	DAF SB220LC550	Ikarus CitiBus	B50F	1992	
20	J24GCX	DAF SB220LC550	Ikarus CitiBus	B50F	1991	Ex Pride of the Road, Royston, 1992
21	J25GCX	DAF SB220LC550	Ikarus CitiBus	B50F	1991	Ex Pride of the Road, Royston, 1992
22	J995GCP	DAF SB220LC550	Ikarus CitiBus	B50F	1991	Ex London Coaches, 1992
23	J996GCP	DAF SB220LC550	Ikarus CitiBus	B50F	1991	Ex Pride of the Road, Royston, 1992
24	J997GCP	DAF SB220LC550	Ikarus CitiBus	B50F	1991	Ex Maidstone, 1992
25	J998GCP	DAF SB220LC550	Ikarus CitiBus	B50F	1991	Ex Pride of the Road, Royston, 1992
31	L540EHD	DAF SB220LC550	Ikarus CitiBus	B50F	1994	
32	L541EHD	DAF SB220LC550	Ikarus CitiBus	B50F	1994	
33	L542EHD	DAF SB220LC550	Ikarus CitiBus	B50F	1994	
34	L543EHD	DAF SB220LC550	Ikarus CitiBus	B50F	1994	
41	HDZ8350	Bova FHD12.290	Bova Futura	C49FT	1989	
43	HDZ8352	Bova FHD12.290	Bova Futura	C49FT	1989	Ex Harris, Catshill, 1990
44	5010CD	Bova FHD12.290	Bova Futura	C32FT	1990	
45	787LOM	Bova FHD12.290	Bova Futura	C40FT	1990	
46	D932ODA	MCW Metroliner DR140/1	MCW 400GT	CH47/16DT	1986	
47	245DOC	DAF SBR3000DKZ570	Plaxton Paramount 4000 III	CH55/19CT	1990	
48	WLT702	DAF SBR3000DKZ570	Plaxton Paramount 4000 III	CH55/19CT	1990	
49	J844RAC	Volvo B10M-60	Ikarus Blue Danube	C53F	1991	
50	J845RAC	Volvo B10M-60	Ikarus Blue Danube	C53F	1991	
51	H407LVC	Volvo B10M-61	Ikarus Blue Danube	C49FT	1991	
52	H408LVC	Volvo B10M-61	Ikarus Blue Danube	C49FT	1991	
53	H130MRW	Volvo B10M-61	Ikarus Blue Danube	C51FT	1991	
54	H131MRW	Volvo B10M-61	Ikarus Blue Danube	C51FT	1991	
55	GIL2942	DAF SB2300DKV601	Van Hool Alizée	C51FT	1988	
57	TIA5734	DAF SB2300DKV601	Van Hool Alizée	C51FT	1988	
58	J58GCX	DAF SB220LC550	Ikarus CitiBus	B50F	1991	
59	G543JOG	Bova FHD12.290	Bova Futura	C46FT	1990	

The Your Bus fleet is maintained as a separate entity to the W M Buses fleet, though both are part of West Midlands Travel, and a subsidiary of National Express. Recently repainted with WM Your Bus titles is 150, H150SKU, seen in Navigation Street, Birmingham. *Tony Wilson*

61	FYX817W	Leyland Leopard PSU3E/4R	Duple Dominant II Express	C49F	1980	
62u	FYX818W	Leyland Leopard PSU3E/4R	Duple Dominant II Express	C49F	1980	
63	N53FWU	DAF DE33WFSB3000	Van Hool Alizée	C51FT	1996	
64	N54FWU	DAF DE33WFSB3000	Van Hool Alizée	C51FT	1996	
101	UHG734R	Leyland National 11351A/1R		B49F	1976	Ex Ribble, 1994
102	CBV785S	Leyland National 11351A/1R		B49F	1977	Ex Ribble, 1994
103	UHG740R	Leyland National 11351A/1R		B49F	1976	Ex Ribble, 1994
104	UHG725R	Leyland National 11351A/1R		B49F	1976	Ex Ribble, 1994
105	NTC623M	Leyland National 1151/1R/0401		B49F	1974	Ex Metrowest, Coseley, 1993
106	NTC619M	Leyland National 1151/1R/0401		B49F	1974	Ex Metrowest, Coseley, 1993
113	PUK633R	Leyland National 11351A/1R(Volvo)		B52F	1979	Ex Midland Red West, 1994
114	PUK634R	Leyland National 11351A/1R(Volvo)		B52F	1979	Ex Midland Red West, 1994
115	NTC606M	Leyland National 1151/1R/0401		B49F	1974	Ex Metrowest, Coseley, 1993
116	NTC609M	Leyland National 1151/1R/0401		B49F	1974	Ex Metrowest, Coseley, 1993
117	NTC612M	Leyland National 1151/1R/0401		B49F	1974	Ex Metrowest, Coseley, 1993
118	AOL17T	Leyland National 11351A/1R		B50F	1979	
141	BOK28V	MCW Metrobus DR102/12	MCW	O43/30F	1980	
142	WDA979T	Leyland Fleetline FE30AGR	MCW	H43/33F	1979	
143	D938NDA	MCW Metrobus DR102/59	MCW	DPH43/30F	1986	
145	SDA628S	Leyland Fleetline FE30AGR	Park Royal	H43/32F	1977	
146	ROX658Y	MCW Metrobus DR102/27	MCW	H43/30F	1983	
147	WDA940T	Leyland Fleetline FE30AGR	MCW	H43/33F	1979	
148	TVP889S	Leyland Fleetline FE30AGR	MCW	H43/32F	1978	

149-159		Volvo B10M-55		Plaxton Derwent		B55F	1990		
149	H149SKU	152	H152SKU	154	H154SKU	156	H156SKU	158	H158SKU
150	H150SKU	153	H153SKU	155	H155SKU	157	H157SKU	159	H159SKU
151	H151SKU								

202	N23FWU	DAF DE02LTSB220	Ikarus CitiBus	B37D	1995	

The North & West Midlands Bus Handbook

Your Bus operate sixteen Ikarus-bodied DAF buses. Pictured while working service 50Y to Druids Heath is one of the type, 21, J25GCX. In recent times this Hungarian product has gained in UK sales partly due to Hughes DAF who have been marketing the product. *Tony Wilson*

W M Buses

511	E511TOV	Iveco Daily 49.10	Carlyle	B25F	1988	
512	E512TOV	Iveco Daily 49.10	Carlyle	B25F	1988	
553	D553NOE	Ford Transit 190D	Carlyle	B18F	1986	
554	D554NOE	Ford Transit 190D	Carlyle	B18F	1986	

566-582
Renault-Dodge S56 — Reeve Burgess — B19F — 1986

566	D566NDA	569	D569NDA	578	D578NDA	581	D581NDA	582	D582NDA
568	D568NDA	575	D575NDA	579	D579NDA				

584-596
Iveco Daily 49.10 — Robin Hood City Nippy — B19F — 1986

584	D584NDA	588	D588NDA	591	D591NDA	593	D593NDA	595	D595NDA
585	D585NDA	589	D589NDA	592	D592NDA	594	D594NDA	596	D596NDA
586	D586NDA								

601-620
MCW MetroRider MF150/3 — MCW — B23F — 1987

601	D601NOE	607	D607NOE	612	D612NOE	615	D615NOE	618	D618NOE
602	D602NOE	608	D608NOE	613	D613NOE	616	D616NOE	619	D619NOE
603	D603NOE	609	D609NOE	614	D614NOE	617	D617NOE	620	D620NOE
605	D605NOE	611	D611NOE						

621-650
MCW MetroRider MF150/4 — MCW — B23F — 1987

621	D621NOE	627	D627NOE	633	D633NOE	640	D640NOE	646	D646NOE
622	D622NOE	628	D628NOE	634	D634NOE	641	D641NOE	647	D647NOE
623	D623NOE	629	D629NOE	636	D635NOE	643	D642NOE	648	D648NOE
624	D624NOE	631	D631NOE	637	D636NOE	644	D643NOE	649	D649NOE
625	D625NOE	632	D632NOE	639	D637NOE	645	D644NOE	650	D650NOE
626	D626NOE								

The North & West Midlands Bus Handbook

4 3 KINGS HEATH
Buses
F671 YOG

257 STOURBRIDGE
via Kingswinford
806
ACCEPTED FREE
Buses
Buses Wolverhampton
DENNIS
K916 FVC

In 1990, West Midlands Travel took delivery of twenty carlyle-bodied Mercedes-Benz 708s, the first minibuses for some two years. Seen crossing over New street rail station is 718, H718LOX.
Tony Wilson

651-665		MCW MetroRider MF150/17	MCW		B23F	1987		
651	E651RVP	654	E654SOL	657	E657RVP	660	E660RVP	663	E663RVP
652	E652RVP	655	E655RVP	658	E658RVP	661	E661RVP	664	E664RVP
653	E653RVP	656	E656RVP	659	E659RVP	662	E662RVP	665	E665RVP

666-685		MCW MetroRider MF150/113	MCW		B23F	1988		
666	F666YOG	670	F670YOG	674	F674YOG	678	F678YOG	682	F682YOG
667	F667YOG	671	F671YOG	675	F675YOG	679	F679YOG	683	F683YOG
668	F668YOG	672	F672YOG	676	F676YOG	680	F680YOG	684	F684YOG
669	F669YOG	673	F673YOG	677	F677YOG	681	F681YOG	685	F685YOG

| 700 | E290OMG | Mercedes-Benz 709D | | Reeve Burgess Beaver | DP25F | 1988 | Ex County, 1995 |

701-720		Mercedes-Benz 709D		Carlyle	B25F	1990		
701	G701HOP	705	G705HOP	709	G709HOP	713	G713HOP	717	G717HOP
702	G702HOP	706	G706HOP	710	G710HOP	714	G714HOP	718	H718LOX
703	G703HOP	707	G707HOP	711	G711HOP	715	G715HOP	719	H719LOX
704	G704HOP	708	G708HOP	712	G712HOP	716	G716HOP	720	H720LOX

Smaller vehicles of W M Buses penetrate into the heart of Birmingham. Based at Central, and seen near New Street is, *opposite, top*, MCW MetroRider 671, F671YOG. Many of the earlier MetroRiders have been placed into the hire fleet and are now elsewhere. *Opposite bottom:* Several of the former YourBus vehicles have lost their orange livery in favour of full W M Buses scheme. Seen in Dudley is 806, K916FVC, a Dennis Dart with Plaxton Pointer bodywork. *Tony Wilson*

The North & West Midlands Bus Handbook

The search for a successor to the Leyland Nationals led to a delivery of six Volvo Citybus YV31s in 1986, a forerunner of the B10M. Fitted with Alexander P-type bodywork they are all based at Wolverhampton where 1059, C59HOM was working when photographed. *Tony Wilson*

801-805

		Dennis Dart 9SDL3011		Wright Handybus		B34F	1991		
801	KDZ5801	802	KDZ5802	803	KDZ5803	804	KDZ5804	805	KDZ5805
806	K916FVC	Dennis Dart 9.8SDL3017		Plaxton Pointer		B40F	1992	Ex Your Bus, 1995	
807	K917FVC	Dennis Dart 9.8SDL3017		Plaxton Pointer		B40F	1992	Ex Your Bus, 1995	
808	K918FVC	Dennis Dart 9.8SDL3017		Plaxton Pointer		B40F	1992	Ex Your Bus, 1995	
809	J348GKH	Dennis Dart 9.8SDL3004		Plaxton Pointer		B40F	1991	Ex Your Bus, 1995	
810	J853TRW	Dennis Dart 9.8SDL3012		Plaxton Pointer		B40F	1992	Ex Your Bus, 1995	
811	J997UAC	Dennis Dart 9.8SDL3012		Plaxton Pointer		B40F	1992	Ex Your Bus, 1995	

1009-1015

		Leyland National 11351A/1R				B50F	1979		
1009	AOL9T	1011	AOL11T	1013	AOL13T	1014	AOL14T	1015	AOL15T
1010	AOL10T	1012	AOL12T						

1021-1047

		Leyland National 2 NL116L11/1R				B50F	1980	1025/33/5/9 are B37D	
1021	DOC21V	1025	DOC25V	1028	DOC28V	1035	DOC35V	1039	DOC39V
1022	DOC22V	1026	DOC26V	1029	DOC29V	1037	DOC37V	1047	DOC47V

1048-1052

		Leyland National 2 NL106L11/1R				B42F	1980	*1048 has a DAF engine	
1048	DOC48V	1049	DOC49V	1050	DOC50V	1051	DOC51V	1052	DOC52V

1053	B53AOC	Dennis Lancet SDA520		Duple Dominant		DP23DL	1985		
1054	B54AOC	Dennis Lancet SDA520		Duple Dominant		DP23DL	1985		

1055-1060

		Volvo Citybus YV31MEC		Alexander P		B50F	1986		
1055	C55HOM	1057	C57HOM	1058	C58HOM	1059	C59HOM	1060	C60HOM
1056	C56HOM								

Following the closure of the Lynx assembly line, the then West Midlands Travel moved to the successor of that model, the Volvo B10B. The first ten to join the fleet had Alexander Strider bodies. Seen in Dudley on the Wolverhampton service is 1324, M324LJW. *Richard Godfrey*

Delivery has now commenced of 200 Wright-bodied Volvo single-decks. Initially ordered as B10B's these will, after the first fifty are delivered, change to the lower B10L model. Photographed with Dudley Castle and Zoo in the background is 1337, N337WOH. *Tony Wilson*

1061-1066 Leyland Lynx LX1126LXCTFR1 Leyland B48F 1986

1061	C61HOM	1063	C63HOM	1064	C64HOM	1065	C65HOM
1062	C62HOM					1066	C66HOM

1067-1316 Leyland Lynx LX2R11C15Z4R Leyland B49F 1989

1067	F67DDA	1117	G117EOG	1167	G167EOG	1217	G217EOG	1267	G267EOG
1068	F68DDA	1118	G118EOG	1168	G168EOG	1218	G218EOG	1268	G268EOG
1069	F69DDA	1119	G119EOG	1169	G169EOG	1219	G219EOG	1269	G269EOG
1070	F70DDA	1120	G120EOG	1170	G170EOG	1220	G220EOG	1270	G270EOG
1071	F71DDA	1121	G121EOG	1171	G171EOG	1221	G221EOG	1271	G271EOG
1072	F72DDA	1122	G122EOG	1172	G172EOG	1222	G222EOG	1272	G272EOG
1073	F73DDA	1123	G123EOG	1173	G173EOG	1223	G223EOG	1273	G273EOG
1074	F74DDA	1124	G124EOG	1174	G174EOG	1224	G224EOG	1274	G274EOG
1075	F75DDA	1125	G125EOG	1175	G175EOG	1225	G225EOG	1275	G275EOG
1076	F76DDA	1126	G126EOG	1176	G176EOG	1226	G226EOG	1276	G276EOG
1077	F77DDA	1127	G127EOG	1177	G177EOG	1227	G227EOG	1277	G277EOG
1078	F78DDA	1128	G128EOG	1178	G178EOG	1228	G228EOG	1278	G278EOG
1079	G79EOG	1129	G129EOG	1179	G179EOG	1229	G229EOG	1279	G279EOG
1080	G80EOG	1130	G130EOG	1180	G180EOG	1230	G230EOG	1280	G280EOG
1081	G81EOG	1131	G131EOG	1181	G181EOG	1231	G231EOG	1281	G281EOG
1082	G82EOG	1132	G132EOG	1182	G182EOG	1232	G232EOG	1282	G282EOG
1083	G83EOG	1133	G133EOG	1183	G183EOG	1233	G233EOG	1283	G283EOG
1084	G84EOG	1134	G134EOG	1184	G184EOG	1234	G234EOG	1284	G284EOG
1085	G85EOG	1135	G135EOG	1185	G185EOG	1235	G235EOG	1285	G285EOG
1086	G86EOG	1136	G136EOG	1186	G186EOG	1236	G236EOG	1286	G286EOG
1087	G87EOG	1137	G137EOG	1187	G187EOG	1237	G237EOG	1287	G287EOG
1088	G88EOG	1138	G138EOG	1188	G188EOG	1238	G238EOG	1288	G288EOG
1089	G89EOG	1139	G139EOG	1189	G189EOG	1239	G239EOG	1289	G289EOG
1090	G90EOG	1140	G140EOG	1190	G190EOG	1240	G240EOG	1290	G290EOG
1091	G91EOG	1141	G141EOG	1191	G191EOG	1241	G241EOG	1291	G291EOG
1092	G92EOG	1142	G142EOG	1192	G192EOG	1242	G242EOG	1292	G292EOG
1093	G93EOG	1143	G143EOG	1193	G193EOG	1243	G243EOG	1293	G293EOG
1094	G94EOG	1144	G144EOG	1194	G194EOG	1244	G244EOG	1294	G294EOG
1095	G95EOG	1145	G145EOG	1195	G195EOG	1245	G245EOG	1295	G295EOG
1096	G96EOG	1146	G146EOG	1196	G196EOG	1246	G246EOG	1296	G296EOG
1097	G97EOG	1147	G147EOG	1197	G197EOG	1247	G247EOG	1297	G297EOG
1098	G98EOG	1148	G148EOG	1198	G198EOG	1248	G248EOG	1298	G298EOG
1099	G99EOG	1149	G149EOG	1199	G199EOG	1249	G249EOG	1299	G299EOG
1100	G100EOG	1150	G150EOG	1200	G200EOG	1250	G250EOG	1300	G300EOG
1101	G101EOG	1151	G151EOG	1201	G201EOG	1251	G251EOG	1301	G301EOG
1102	G102EOG	1152	G152EOG	1202	G202EOG	1252	G252EOG	1302	G302EOG
1103	G103EOG	1153	G153EOG	1203	G203EOG	1253	G253EOG	1303	G303EOG
1104	G104EOG	1154	G154EOG	1204	G204EOG	1254	G254EOG	1304	G304EOG
1105	G105EOG	1155	G155EOG	1205	G205EOG	1255	G255EOG	1305	G305EOG
1106	G106EOG	1156	G156EOG	1206	G206EOG	1256	G256EOG	1306	G306EOG
1107	G107EOG	1157	G157EOG	1207	G207EOG	1257	G257EOG	1307	G307EOG
1108	G108EOG	1158	G158EOG	1208	G208EOG	1258	G258EOG	1308	G308EOG
1109	G109EOG	1159	G159EOG	1209	G209EOG	1259	G259EOG	1309	G309EOG
1110	G110EOG	1160	G160EOG	1210	G210EOG	1260	G260EOG	1310	G310EOG
1111	G111EOG	1161	G161EOG	1211	G211EOG	1261	G261EOG	1311	G311EOG
1112	G112EOG	1162	G162EOG	1212	G212EOG	1262	G262EOG	1312	G312EOG
1113	G113EOG	1163	G163EOG	1213	G213EOG	1263	G263EOG	1313	G313EOG
1114	G114EOG	1164	G164EOG	1214	G214EOG	1264	G264EOG	1314	G314EOG
1115	G115EOG	1165	G165EOG	1215	G215EOG	1265	G265EOG	1315	G315EOG
1116	G116EOG	1166	G166EOG	1216	G216EOG	1266	G266EOG	1316	G316EOG

1317-1326 Volvo B10B Alexander Strider B51F 1994

1317	M317LJW	1319	M319LJW	1321	M321LJW	1323	M323LJW	1325	M325LJW
1318	M318LJW	1320	M320LJW	1322	M322LJW	1324	M324LJW	1326	M326LJW

| 1327 | M845OKV | Volvo B10B | | Wright Endurance | B51F | 1995 | On loan from Volvo | |
| 1328 | N986TWK | Volvo B10B | | Wright Endurance | B51F | 1995 | On loan from Volvo | |

1329-1378

Volvo B10B — Wright Endurance — B48F — 1995

1329	N329WOH	1339	N339WOH	1349	N349WOH	1359	N359WOH	1369	N369WOH
1330	N330WOH	1340	N340WOH	1350	N350WOH	1360	N360WOH	1370	N370WOH
1331	N331WOH	1341	N341WOH	1351	N351WOH	1361	N361WOH	1371	N371WOH
1332	N332WOH	1342	N342WOH	1352	N352WOH	1362	N362WOH	1372	N372WOH
1333	N133WOH	1343	N343WOH	1353	N353WOH	1363	N363WOH	1373	N373WOH
1334	N334WOH	1344	N344WOH	1354	N354WOH	1364	N364WOH	1374	N374WOH
1335	N335WOH	1345	N345WOH	1355	N355WOH	1365	N365WOH	1375	N375WOH
1336	N336WOH	1346	N346WOH	1356	N356WOH	1366	N366WOH	1376	N376WOH
1337	N337WOH	1347	N347WOH	1357	N357WOH	1367	N367WOH	1377	N377WOH
1338	N338WOH	1348	N348WOH	1358	N358WOH	1368	N368WOH	1378	N378WOH

1470-1518

Leyland National 11351/1R (DAF) — B50F — 1974

1470	ROK470M	1478	TOE478N	1499	TOE499N	1507	TOE507N	1517	TOE517N
1477	TOE477N	1485	TOE485N	1502	TOE502N	1509	TOE509N	1518	GOK518N

1631	STJ31T	Leyland National 11351A/1R(Volvo)	B52F	1979	
1648	LMB948P	Leyland National 11351/1R(Volvo)	B50F	1975	
1659	SGR559R	Leyland National 11351A/1R(Volvo)	B52F	1976	Ex Metrowest, Coseley, 1993
1662	LJN622P	Leyland National 11351/1R(Volvo)	B49F	1975	Ex Metrowest, Coseley, 1993
1731	VNO731S	Leyland National 11351A/1R(Volvo)	B49F	1977	Ex Metrowest, Coseley, 1993
1734	STJ34T	Leyland National 11351A/1R(Volvo)	B52F	1979	Ex Metrowest, Coseley, 1993
1745	PTF745L	Leyland National 1151/2R/0402(Volvo)	B52F	1973	
1758	PTF758L	Leyland National 1151/1R(DAF)	B52F	1973	Ex Metrowest, Coseley, 1993

1802-1825

Leyland National 11351A/1R (DAF)* — B50F* — 1977 — *1821/5 are DP45F
*1821/5 have Volvo engines

1802	OOX802R	1806	OOX806R	1810	OOX810R	1818	OOX818R	1822	OOX822R
1804	OOX804R	1807	OOX807R	1812	OOX812R	1821	OOX821R	1825	OOX825R
1805	OOX805R	1808	OOX808R	1813	OOX813R				

1836-1850

Leyland National 11351A/1R(DAF) — B50F* — 1978 — *1847 is B22DL

1836	TVP836S	1839	TVP839S	1842	TVP842S	1846	TVP846S	1848	TVP848S
1837	TVP837S	1840	TVP840S	1843	TVP843S	1847	TVP847S	1850	TVP850S
1838	TVP838S	1841	TVP841S						

1854-1865

Leyland National 11351A/1R (Volvo) — DP45F — 1978

1854	TVP854S	1862	TVP862S	1863	TVP863S	1864	TVP864S	1865	TVP865S
1856	TVP856S								

1956	WDA956T	Leyland Fleetline FE30AGR	MCW/WMT	B37F	1978

Leyland Nationals are being displaced quite quickly from the W M Buses fleet, even though they received many improvements during the time with the company. Seen in Wolverhampton is 1010, AOL10T, one of the 1979 batch.
Phillip Stephenson

2001-2074   MCW Metrobus DR102/12   MCW   H43/30F 1979-80

2001	BOK1V	2016	BOK16V	2032	BOK32V	2046	BOK46V	2060	BOK60V
2002	BOK2V	2017	BOK17V	2033	BOK33V	2047	BOK47V	2061	BOK61V
2003	BOK3V	2018	BOK18V	2034	BOK34V	2048	BOK48V	2062	BOK62V
2005	BOK5V	2019	BOK19V	2035	BOK35V	2049	BOK49V	2063	BOK63V
2006	BOK6V	2020	BOK20V	2036	BOK36V	2050	BOK50V	2064	BOK64V
2007	BOK7V	2021	BOK21V	2037	BOK37V	2051	BOK51V	2065	BOK65V
2008	BOK8V	2022	BOK22V	2038w	BOK38V	2052	BOK52V	2066	BOK66V
2009	BOK9V	2023	BOK23V	2039	BOK39V	2053	BOK53V	2067	BOK67V
2010	BOK10V	2025	BOK25V	2040	BOK40V	2054	BOK54V	2069	BOK69V
2011	BOK11V	2026	BOK26V	2041	BOK41V	2055	BOK55V	2070	BOK70V
2012	BOK12V	2027	BOK27V	2042	BOK42V	2056	BOK56V	2071	BOK71V
2013	BOK13V	2028	BOK28V	2043	BOK43V	2057	BOK57V	2073	BOK73V
2014	BOK14V	2030	BOK30V	2044	BOK44V	2058	BOK58V	2074	BOK74V
2015	BOK15V	2031	BOK31V	2045	BOK45V	2059	BOK59V		

2076-2225   MCW Metrobus DR102/18   MCW   H43/30F 1980-81

2076	BOK76V	2106	GOG106W	2136	GOG136W	2167	GOG167W	2196	GOG196W
2077	BOK77V	2107	GOG107W	2137	GOG137W	2168	GOG168W	2197	GOG197W
2078	BOK78V	2108	GOG108W	2138	GOG138W	2169	GOG169W	2198	GOG198W
2079	BOK79V	2109	GOG109W	2139	GOG139W	2170	GOG170W	2199	GOG199W
2080	BOK80V	2110	GOG110W	2140	GOG140W	2171	GOG171W	2200	GOG200W
2081	BOK81V	2111	GOG111W	2141	GOG141W	2172	GOG172W	2201	GOG201W
2082	BOK82V	2112	GOG112W	2142	GOG142W	2173	GOG173W	2202	GOG202W
2083	BOK83V	2113	GOG113W	2143	GOG143W	2174	GOG174W	2203	GOG203W
2084	BOK84V	2114	GOG114W	2144	GOG144W	2175	GOG175W	2204	GOG204W
2085	BOK85V	2115	GOG115W	2145	GOG145W	2176	GOG176W	2205	GOG205W
2086	BOK86V	2116	GOG116W	2146	GOG146W	2177	GOG177W	2206	GOG206W
2087	BOK87V	2117	GOG117W	2147	GOG147W	2178	GOG178W	2207	GOG207W
2088	BOK88V	2118	GOG118W	2148	GOG148W	2179	GOG179W	2208	GOG208W
2089	BOK89V	2119	GOG119W	2149	GOG149W	2180	GOG180W	2209	GOG209W
2090	BOK90V	2120	GOG120W	2150	GOG150W	2181	GOG181W	2210	GOG210W
2091	GOG91W	2121	GOG121W	2151	GOG151W	2182	GOG182W	2211	GOG211W
2092	GOG92W	2122	GOG122W	2152	GOG152W	2183	GOG183W	2212	GOG212W
2093	GOG93W	2123	GOG123W	2153	GOG153W	2184	GOG184W	2213	GOG213W
2094	GOG94W	2124	GOG124W	2155	GOG155W	2185	GOG185W	2214	GOG214W
2095	GOG95W	2125	GOG125W	2156	GOG156W	2186	GOG186W	2215	GOG215W
2096	GOG96W	2126	GOG126W	2157	GOG157W	2187	GOG187W	2216	GOG216W
2097	GOG97W	2127	GOG127W	2158	GOG158W	2188	GOG188W	2217	GOG217W
2098	GOG98W	2128	GOG128W	2159	GOG159W	2189	GOG189W	2218	GOG218W
2099	GOG99W	2129	GOG129W	2161	GOG161W	2190	GOG190W	2219	GOG219W
2100	GOG100W	2130	GOG130W	2162	GOG162W	2191	GOG191W	2220	GOG220W
2101	GOG101W	2131	GOG131W	2163	GOG163W	2192	GOG192W	2221	GOG221W
2102	GOG102W	2132	GOG132W	2164	GOG164W	2193	GOG193W	2222	GOG222W
2103	GOG103W	2133	GOG133W	2165	GOG165W	2194	GOG194W	2224	GOG224W
2104	GOG104W	2134	GOG134W	2166	GOG166W	2195	GOG195W	2225	GOG225W
2105	GOG105W	2135	GOG135W						

2226-2245   MCW Metrobus DR104/8   MCW   H43/30F 1981

2226	GOG226W	2230	GOG230W	2234	GOG234W	2238	GOG238W	2242	GOG242W
2227	GOG227W	2231	GOG231W	2235	GOG235W	2239	GOG239W	2243	GOG243W
2228	GOG228W	2232	GOG232W	2236	GOG236W	2240	GOG240W	2244	GOG244W
2229	GOG229W	2233	GOG233W	2237	GOG237W	2241	GOG241W	2245	GOG245W

2246-2275   MCW Metrobus DR102/18   MCW   H43/30F 1981

2246	GOG246W	2252	GOG252W	2258	GOG258W	2264	GOG264W	2270	GOG270W
2247	GOG247W	2253	GOG253W	2259	GOG259W	2265	GOG265W	2271	GOG271W
2248	GOG248W	2254	GOG254W	2260	GOG260W	2266	GOG266W	2273	GOG273W
2249	GOG249W	2255	GOG255W	2261	GOG261W	2267	GOG267W	2274	GOG274W
2250	GOG250W	2256	GOG256W	2262	GOG262W	2268	GOG268W	2275	GOG275W
2251	GOG251W	2257	GOG257W	2263	GOG263W	2269	GOG269W		

2276-2325   MCW Metrobus DR102/22   MCW   H43/30F   1981

2276	KJW276W	2284	KJW284W	2292	KJW292W	2302	KJW302W	2315	KJW315W
2277	KJW277W	2285	KJW285W	2293	KJW293W	2303	KJW303W	2316	KJW316W
2278	KJW278W	2286	KJW286W	2294	KJW294W	2308	KJW308W	2317	KJW317W
2279	KJW279W	2287	KJW287W	2295	KJW295W	2309	KJW309W	2319	KJW319W
2280	KJW280W	2288	KJW288W	2297	KJW297W	2311	KJW311W	2321	KJW321W
2281	KJW281W	2289	KJW289W	2298	KJW298W	2312	KJW312W	2323	KJW323W
2282	KJW282W	2290	KJW290W	2299	KJW299W	2313	KJW313W	2324	KJW324W
2283	KJW283W	2291	KJW291W	2300	KJW300W	2314	KJW314W	2325	KJW325W

2326-2435   MCW Metrobus DR102/22   MCW   H43/30F   1981-82

2326	LOA326X	2348	LOA348X	2370	LOA370X	2392	LOA392X	2414	LOA414X
2327	LOA327X	2349	LOA349X	2371	LOA371X	2393	LOA393X	2415	LOA415X
2328	LOA328X	2350	LOA350X	2372	LOA372X	2394	LOA394X	2416	LOA416X
2329	LOA329X	2351	LOA351X	2373	LOA373X	2395	LOA395X	2417	LOA417X
2330	LOA330X	2352	LOA352X	2374	LOA374X	2396	LOA396X	2418	LOA418X
2331	LOA331X	2353	LOA353X	2375	LOA375X	2397	LOA397X	2419	LOA419X
2332	LOA332X	2354	LOA354X	2376	LOA376X	2398	LOA398X	2420	LOA420X
2333	LOA333X	2355	LOA355X	2377	LOA377X	2399	LOA399X	2421	LOA421X
2334	LOA334X	2356	LOA356X	2378	LOA378X	2400	LOA400X	2422	LOA422X
2335	LOA335X	2357	LOA357X	2379	LOA379X	2401	LOA401X	2423	LOA423X
2336	LOA336X	2358	LOA358X	2380	LOA380X	2402	LOA402X	2424	LOA424X
2337	LOA337X	2359	LOA359X	2381	LOA381X	2403	LOA403X	2425	LOA425X
2338	LOA338X	2360	LOA360X	2382	LOA382X	2404	LOA404X	2426	LOA426X
2339	LOA339X	2361	LOA361X	2383	LOA383X	2405	LOA405X	2427	LOA427X
2340	LOA340X	2362	LOA362X	2384	LOA384X	2406	LOA406X	2428	LOA428X
2341	LOA341X	2363	LOA363X	2385	LOA385X	2407	LOA407X	2429w	LOA429X
2342	LOA342X	2364	LOA364X	2386	LOA386X	2408	LOA408X	2430	LOA430X
2343	LOA343X	2365	LOA365X	2387	LOA387X	2409	LOA409X	2431	LOA431X
2344	LOA344X	2366	LOA366X	2388w	LOA388X	2410	LOA410X	2432	LOA432X
2345	LOA345X	2367	LOA367X	2389	LOA389X	2411	LOA411X	2433	LOA433X
2346	LOA346X	2368	LOA368X	2390	LOA390X	2412	LOA412X	2434	LOA434X
2347	LOA347X	2369	LOA369X	2391	LOA391X	2413	LOA413X	2435	LOA435X

2436-2475   MCW Metrobus DR102/27   MCW   H43/30F   1982

2436	NOA436X	2444	NOA444X	2452	NOA452X	2460	NOA460X	2468	NOA468X
2437	NOA437X	2445	NOA445X	2453	NOA453X	2461	NOA461X	2469	NOA469X
2438	NOA438X	2446	NOA446X	2454	NOA454X	2462	NOA462X	2470	NOA470X
2439	NOA439X	2447	NOA447X	2455	NOA455X	2463	NOA463X	2471	NOA471X
2440	NOA440X	2448	NOA448X	2456	NOA456X	2464	NOA464X	2472	NOA472X
2441	NOA441X	2449	NOA449X	2457	NOA457X	2465	NOA465X	2473	NOA473X
2442	NOA442X	2450	NOA450X	2458	NOA458X	2466	NOA466X	2474	NOA474X
2443	NOA443X	2451	NOA451X	2459	NOA459X	2467	NOA467X	2475	NOA475X

One of West Bromwich's allocation of Metrobuses is 2221 seen heading for Bearwood. The Metrobuses have now commenced a refurbishment programme.
Phillip Stephenson

2476-2610 MCW Metrobus DR102/27 MCW H43/30F 1982-83

2476	POG476Y	2503	POG503Y	2530	POG530Y	2557	POG557Y	2584	POG584Y
2477	POG477Y	2504	POG504Y	2531	POG531Y	2558	POG558Y	2585	POG585Y
2478	POG478Y	2505	POG505Y	2532	POG532Y	2559	POG559Y	2586	POG586Y
2479	POG479Y	2506	POG506Y	2533	POG533Y	2560	POG560Y	2587	POG587Y
2480	POG480Y	2507	POG507Y	2534	POG534Y	2561	POG561Y	2588	POG588Y
2481	POG481Y	2508	POG508Y	2535	POG535Y	2562	POG562Y	2589	POG589Y
2482	POG482Y	2509	POG509Y	2536	POG536Y	2563	POG563Y	2590	POG590Y
2483	POG483Y	2510	POG510Y	2537	POG537Y	2564	POG564Y	2591	POG591Y
2484	POG484Y	2511	POG511Y	2538	POG538Y	2565	POG565Y	2592	POG592Y
2485	POG485Y	2512	POG512Y	2539	POG539Y	2566	POG566Y	2593	POG593Y
2486	POG486Y	2513	POG513Y	2540	POG540Y	2567	POG567Y	2594	POG594Y
2487	POG487Y	2514	POG514Y	2541	POG541Y	2568	POG568Y	2595	POG595Y
2488	POG488Y	2515	POG515Y	2542	POG542Y	2569	POG569Y	2596	POG596Y
2489	POG489Y	2516	POG516Y	2543	POG543Y	2570	POG570Y	2597	POG597Y
2490	POG490Y	2517	POG517Y	2544	POG544Y	2571	POG571Y	2598	POG598Y
2491	POG491Y	2518	POG518Y	2545	POG545Y	2572	POG572Y	2599	POG599Y
2492	POG492Y	2519	POG519Y	2546	POG546Y	2573	POG573Y	2600	POG600Y
2493	POG493Y	2520	POG520Y	2547	POG547Y	2574	POG574Y	2601	POG601Y
2494	POG494Y	2521	POG521Y	2548	POG548Y	2575	POG575Y	2602	POG602Y
2495	POG495Y	2522	POG522Y	2549	POG549Y	2576	POG576Y	2603	POG603Y
2496	POG496Y	2523	POG523Y	2550	POG550Y	2577	POG577Y	2604	POG604Y
2497	POG497Y	2524	POG524Y	2551	POG551Y	2578	POG578Y	2605	POG605Y
2498	POG498Y	2525	POG525Y	2552	POG552Y	2579	POG579Y	2606	POG606Y
2499	POG499Y	2526	POG526Y	2553	POG553Y	2580	POG580Y	2607	POG607Y
2500	POG500Y	2527	POG527Y	2554	POG554Y	2581	POG581Y	2608	POG608Y
2501	POG501Y	2528	POG528Y	2555	POG555Y	2582	POG582Y	2609	POG609Y
2502	POG502Y	2529	POG529Y	2556	POG556Y	2583	POG583Y	2610	POG610Y

2611-2667 MCW Metrobus DR102/27 MCW H43/30F 1983

2611	ROX611Y	2623	ROX623Y	2634	ROX634Y	2645	ROX645Y	2656	ROX656Y
2612	ROX612Y	2624	ROX624Y	2635	ROX635Y	2646	ROX646Y	2657	ROX657Y
2613	ROX613Y	2625	ROX625Y	2636	ROX636Y	2647	ROX647Y	146	ROX658Y
2614	ROX614Y	2626	ROX626Y	2637	ROX637Y	2648	ROX648Y	2659	ROX659Y
2615	ROX615Y	2627	ROX627Y	2638	ROX638Y	2649	ROX649Y	2660	ROX660Y
2616	ROX616Y	2628	ROX628Y	2639	ROX639Y	2650	ROX650Y	2661	ROX661Y
2617	ROX617Y	2629	ROX629Y	2640	ROX640Y	2651	ROX651Y	2663	ROX663Y
2618	ROX618Y	2630	ROX630Y	2641	ROX641Y	2652	ROX652Y	2664	ROX664Y
2619	ROX619Y	2631	ROX631Y	2642	ROX642Y	2653	ROX653Y	2665	ROX665Y
2620	ROX620Y	2632	ROX632Y	2643	ROX643Y	2654	ROX654Y	2666	ROX666Y
2621	ROX621Y	2633	ROX633Y	2644	ROX644Y	2655	ROX655Y	2667	ROX667Y
2622	ROX622Y								

2668-2735 MCW Metrobus DR102/27 MCW H43/30F 1983-84

2668	A668UOE	2682	A682UOE	2696	A696UOE	2709	A709UOE	2723	A723UOE
2669	A669UOE	2683	A683UOE	2697	A697UOE	2710	A710UOE	2724	A724UOE
2670	A670UOE	2684	A684UOE	2698	A698UOE	2712	A712UOE	2725	A725UOE
2671	A671UOE	2685	A685UOE	2699	A699UOE	2713	A713UOE	2726	A726UOE
2672	A672UOE	2686	A686UOE	2700	A690UOE	2714	A714UOE	2727	A727UOE
2673	A673UOE	2687	A687UOE	2701	A701UOE	2715	A715UOE	2728	A728UOE
2674	A674UOE	2688	A688UOE	2702	A702UOE	2716	A716UOE	2729	A729UOE
2675	A675UOE	2689	A689UOE	2703	A703UOE	2717	A717UOE	2730	A730UOE
2676	A676UOE	2690	A690UOE	2704	A704UOE	2718	A718UOE	2731	A731UOE
2677	A677UOE	2691	A691UOE	2705	A705UOE	2719	A719UOE	2732	A732UOE
2678	A678UOE	2692	A692UOE	2706	A706UOE	2720	A720UOE	2733	A733UOE
2679	A679UOE	2693	A693UOE	2707	A707UOE	2721	A721UOE	2734	A734UOE
2680	A680UOE	2694	A694UOE	2708	A708UOE	2722	A722UOE	2735	A735UOE
2681	A681UOE	2695	A695UOE						

Newly re-painted and seen in Carrs Lane, Birmingham is Metrobus Mark 2 2594 lettered for the Perry Barr depot. *Tony Wilson*

2736-2772

MCW Metrobus DR102/27   MCW   H43/30F   1984

2736	A736WVP	2744	A744WVP	2752	A752WVP	2759	A759WVP	2766	A766WVP
2737	A737WVP	2745	A745WVP	2753	A753WVP	2760	A760WVP	2767	A767WVP
2738	A738WVP	2746	A746WVP	2754	A754WVP	2761	A761WVP	2768	A768WVP
2739	A739WVP	2747	A747WVP	2755	A755WVP	2762	A762WVP	2769	A769WVP
2740	A740WVP	2748	A748WVP	2756	A756WVP	2763	A763WVP	2770	A770WVP
2741	A741WVP	2749	A749WVP	2757	A757WVP	2764	A764WVP	2771	A771WVP
2742	A742WVP	2750	A750WVP	2758	A758WVP	2765	A765WVP	2772	A772WVP
2743	A743WVP	2751	A751WVP						

2774-2860

MCW Metrobus DR102/27   MCW   H43/30F   1984-85

2774	B774AOC	2792	B792AOC	2810	B810AOP	2827	B827AOP	2844	B844AOP
2775	B775AOC	2793	B793AOC	2811	B811AOP	2828	B828AOP	2845	B845AOP
2776	B776AOC	2794	B794AOC	2812	B812AOP	2829	B829AOP	2846	B846AOP
2777	B777AOC	2795	B795AOC	2813	B813AOP	2830	B830AOP	2847	B847AOP
2778	B778AOC	2796	B796AOC	2814	B814AOP	2831	B831AOP	2848	B848AOP
2779	B779AOC	2797	B797AOP	2815	B815AOP	2832	B832AOP	2849	B849AOP
2780	B780AOC	2798	B798AOP	2816	B816AOP	2833	B833AOP	2850	B850AOP
2781	B781AOC	2799	B799AOP	2817	B817AOP	2834	B834AOP	2851	B851AOP
2782	B782AOC	2800	B800AOP	2818	B818AOP	2835	B835AOP	2852	B852AOP
2783	B783AOC	2801	B801AOP	2819	B819AOP	2836	B836AOP	2853	B853AOP
2784	B784AOC	2802	B802AOP	2820	B820AOP	2837	B837AOP	2854	B854AOP
2785	B785AOC	2803	B803AOP	2821	B821AOP	2838	B838AOP	2855	B855AOP
2786	B786AOC	2804	B804AOP	2822	B822AOP	2839	B839AOP	2856	B856AOP
2787	B787AOC	2805	B805AOP	2823	B823AOP	2840	B840AOP	2857	B857AOP
2788	B788AOC	2806	B806AOP	2824	B824AOP	2841	B841AOP	2858	B858AOP
2789	B789AOC	2807	B807AOP	2825	B825AOP	2842	B842AOP	2859	B859AOP
2790	B790AOC	2808	B808AOP	2826	B826AOP	2843	B843AOP	2860	B860AOP
2791	B791AOC	2809	B809AOP						

Vintage operational vehicles are still popular with many operators who use them for specific duties. W M Buses have re-introduced 3225, MOF225 into the operational fleet. The Daimler CVG6 features a Crossley body a type popular in the municipal fleets of the late 1950s. *Phillip Stephenson*

2861-2910   MCW Metrobus DR102/48   MCW   H43/30F   1985

2861	B861DOM	2871	B871DOM	2881	B881DOM	2891	C891FON	2901	C901FON
2862	B862DOM	2872	B872DOM	2882	B882DOM	2892	C892FON	2902	C902FON
2863	B863DOM	2873	B873DOM	2883	B883DOM	2893	C893FON	2903	C903FON
2864	B864DOM	2874	B874DOM	2884	B884DOM	2894	C894FON	2904	C904FON
2865	B865DOM	2875	B875DOM	2885	B885DOM	2895	C895FON	2905	C905FON
2866	B866DOM	2876	B876DOM	2886	B886DOM	2896	C896FON	2906	C906FON
2867	B867DOM	2877	B877DOM	2887	C887FON	2897	C897FON	2907	C907FON
2868	B868DOM	2878	B878DOM	2888	C888FON	2898	C898FON	2908	C908FON
2869	B869DOM	2879	B879DOM	2889	C889FON	2899	C899FON	2909	C909FON
2870	B870DOM	2880	B880DOM	2890	C890FON	2900	C900FON	2910	C910FON

2911-2960   MCW Metrobus DR102/59   MCW   DPH43/27F   1986

2954 is H43/30F;

2911	D911NDA	2921	D921NDA	2931	D931NDA	2941	D941NDA	2951	D951NDA
2912	D912NDA	2922	D922NDA	2932	D932NDA	2942	D942NDA	2952	D952NDA
2913	D913NDA	2923	D923NDA	2933	D933NDA	2943	D943NDA	2953	D953NDA
2914	D914NDA	2924	D924NDA	2934	D934NDA	2944	D944NDA	2954	D954NDA
2915	D915NDA	2925	D925NDA	2935	D935NDA	2945	D945NDA	2955	D955NDA
2916	D916NDA	2926	D926NDA	2936	D936NDA	2946	D946NDA	2956	D956NDA
2917	D917NDA	2927	D927NDA	2937	D937NDA	2947	D947NDA	2957	D957NDA
2918	D918NDA	2928	D928NDA	143	D938NDA	2948	D948NDA	2958	D958NDA
2919	D919NDA	2929	D929NDA	2939	D939NDA	2949	D949NDA	2959	D959NDA
2920	D920NDA	2930	D930NDA	2940	D940NDA	2950	D950NDA	2960	D960NDA

Opposite, top: Nostalgic liveries for W M Buses fleets were initiated at the end on 1995 with 3030, F30XOF being one of the first examples seen here in Birmingham Corporation's colours. One vehicle at each depot is being painted in appropriate colours for the depot. *G J Kelland*
Opposite, bottom: Representing the forty Scania double-deck buses from 1990 is 3221, H221LOM. The witholding at that time of many 'meaningful' numbers by the DVLA caused the number range to be spread to fleet number 3247. *Tony Wilson*

11A
OUTER CIRCLE
Perry Barr Acocks Green
Kings Heath Bearwood
3030
F30 XOF

CHURCH
50
DRUIDS HEATH
via Moseley Road
3221
Buses Central
H221 LOM

2961-2974　　MCW Metrobus GR133/1　　MCW　　H43/30F　1984

2961	A101WVP	2965	A105WVP	2968	A108WVP	2971	A111WVP	2973	A113WVP
2962	A102WVP	2966	A106WVP	2969	A109WVP	2972	A112WVP	2974	A114WVP
2963	A103WVP	2967	A107WVP	2970	A110WVP				

2975-3046　　MCW Metrobus DR102/64　　MCW　　H43/30F　1988-89

2975	E975VUK	2990	F990XOE	3005	F305XOF	3019	F319XOF	3033	F33XOF
2976	E976VUK	2991	F991XOE	3006	F306XOF	3020	F320XOF	3034	F34XOF
2977	E977VUK	2992	F992XOE	3007	F307XOF	3021	F321XOF	3035	F35XOF
2978	E978VUK	2993	F993XOE	3008	F308XOF	3022	F22XOF	3036	F36XOF
2979	E979VUK	2994	F994XOE	3009	F309XOF	3023	F23XOF	3037	F37XOF
2980	E980VUK	2995	F995XOE	3010	F310XOF	3024	F24XOF	3038	F38XOF
2981	E981VUK	2996	F996XOE	3011	F311XOF	3025	F25XOF	3039	F39XOF
2982	E982VUK	2997	F997XOE	3012	F312XOF	3026	F26XOF	3040	F40XOF
2983	E983VUK	2998	F998XOE	3013	F313XOF	3027	F27XOF	3041	F41XOF
2984	E984VUK	2999	F999XOE	3014	F314XOF	3028	F28XOF	3042	F42XOF
2985	E985VUK	3000	F300XOF	3015	F315XOF	3029	F29XOF	3043	F43XOF
2986	E986VUK	3001	F301XOF	3016	F316XOF	3030	F30XOF	3044	F44XOF
2987	E987VUK	3002	F302XOF	3017	F317XOF	3031	F31XOF	3045	F45XOF
2988	E988VUK	3003	F303XOF	3018	F318XOF	3032	F32XOF	3046	F46XOF
2989	E989VUK	3004	F304XOF						

3047-3124　　MCW Metrobus DR102/70　　MCW　　H43/30F　1989-90

3047	F47XOF	3063	F63XOF	3079	F79XOF	3094	F94XOF	3110	G110FJW
3048	F48XOF	3064	F64XOF	3080	F80XOF	3095	F95XOF	3111	G111FJW
3049	F49XOF	3065	F65XOF	3081	F81XOF	3096	F96XOF	3112	G112FJW
3050	F50XOF	3066	F66XOF	3082	F82XOF	3097	F97XOF	3113	G113FJW
3051	F51XOF	3067	F67XOF	3083	F83XOF	3098	F98XOF	3114	G114FJW
3052	F52XOF	3068	F68XOF	3084	F84XOF	3099	F99XOF	3115	G115FJW
3053	F53XOF	3069	F69XOF	3085	F85XOF	3100	F100XOF	3116	G116FJW
3054	F54XOF	3070	F70XOF	3086	F86XOF	3101	F101XOF	3117	G117FJW
3055	F55XOF	3071	F71XOF	3087	F87XOF	3102	F102XOF	3118	G118FJW
3056	F56XOF	3072	F72XOF	3088	F88XOF	3103	F103XOF	3119	G119FJW
3057	F57XOF	3073	F73XOF	3089	F89XOF	3104	G104FJW	3120	G120FJW
3058	F58XOF	3074	F74XOF	3090	F90XOF	3105	G105FJW	3121	G121FJW
3059	F59XOF	3075	F75XOF	3091	F91XOF	3106	G106FJW	3122	G122FJW
3060	F50XOF	3076	F76XOF	3092	F92XOF	3108	G108FJW	3123	G123FJW
3061	F61XOF	3077	F77XOF	3093	F93XOF	3109	G109FJW	3124	G124FJW
3062	F62XOF	3078	F78XOF						

3201-3247　　Scania N113DRB　　Alexander RH　　H47/33F　1990

3201	H201LOM	3210	H210LOM	3221	H221LOM	3231	H231LOM	3239	H239LOM
3202	H202LOM	3211	H211LOM	3223	H223LOM	3232	H232LOM	3241	H241LOM
3203	H203LOM	3212	H212LOM	3224	H224LOM	3233	H233LOM	3242	H242LOM
3204	H204LOM	3215	H215LOM	3225	H225LOM	3234	H234LOM	3243	H243LOM
3206	H206LOM	3217	H217LOM	3226	H226LOM	3235	H235LOM	3244	H244LOM
3207	H207LOM	3218	H218LOM	3227	H227LOM	3236	H236LOM	3245	H245LOM
3208	H208LOM	3219	H219LOM	3228	H228LOM	3237	H237LOM	3246	H246LOM
3209	H209LOM	3220	H220LOM	3229	H229LOM	3238	H238LOM	3247	H247LOM

3225	MOF225	Daimler CVG6		Crossley	H30/25R	1954 Ex Birmingham, 1969
4069	YOK69K	Daimler Fleetline CRG6LX		Park Royal	O47/30F	1971

6443-6488　　Leyland Fleetline FE30AGR　　MCW　　H47/33F　1976-77

6443	NOC434R	6445	NOC445R	6471	NOC471R	6477	NOC477R	6488	NOC488R
6444	NOC444R								

6515-6557　　Leyland Fleetline FE30AGR　　MCW　　H47/33F　1977-78

6515	SDA515S	6538	SDA538S	6545	SDA545S	6550	SDA550S	6557	SDA557S
6537	SDA537S								

Opposite: **Comparison of Fleetlines are seen here with bodywork by East Lancashire (6741, NOC71R) and MCW (6932, WDA932T).** *Tony Wilson*

It's Nice 'n Easy to be natural.
West Midlands Buses
395
BURNTWOOD
via Brownhills
6741
NOC 741R

West Midlands Buses
55
CHELMSLEY WOOD
via Alum Rock
6932
WDA 932T

6600-6624   Leyland Fleetline FE30AGR   Park Royal   H43/33F   1976-77

6600	NOC600R								
6610	NOC610R	6615	NOC615R	6619	SDA619S	6621	SDA621S	6624	SDA624S

6630-6690   Leyland Fleetline FE30AGR   Park Royal   H43/33F   1977-79

6630	SDA630S	6648	SDA648S	6663	WDA663T	6670	WDA670T	6682	WDA682T
6634	SDA634S	6649	SDA649S	6665	WDA665T	6672	WDA672T	6686	WDA686T
6639	SDA639S	6650	SDA650S	6666	WDA666T	6673	WDA673T	6688	WDA688T
6642	SDA642S	6660	SDA660S	6667	WDA667T	6677	WDA677T	6689	WDA689T
6643	SDA643S	6661	WDA661T	6669	WDA669T	6681	WDA681T	6690	WDA690T
6646	SDA646S	6662	WDA662T						

6699-6718   Leyland Fleetline FE30AGR   MCW   H43/33F   1978

6699	SDA699S	6710	SDA710S	6712	SDA712S	6714	SDA714S	6718	SDA718S
6709	SDA709S								

6721-6760   Leyland Fleetline FE30AGR   East Lancashire   H43/33F   1977-78

6723	NOC723R	6732	NOC732R	6741	NOC741R	6752	SDA752S	6760	SDA760S
6728	NOC728R	6735	NOC735R	6744	NOC744R	6757	SDA757S		

6764	SDA764S	Leyland Fleetline FE30AGR	MCW	H43/33F	1978
6767	SDA767S	Leyland Fleetline FE30AGR	MCW	H43/33F	1978
6772	SDA772S	Leyland Fleetline FE30AGR	MCW	H43/33F	1978
6800	SDA800S	Leyland Fleetline FE30AGR	MCW	H43/33F	1978
6835	WDA835T	MCW Metrobus DR102/1	MCW	H43/30F	1978

6866-6904   Leyland Fleetline FE30AGR   MCW   H43/33F   1978

6866	TVP866S	6874	TVP874S	6885	TVP885S	6895	TVP895S	6901	TVP901S
6871	TVP871S	6875	TVP875S	6888	TVP888S	6897	TVP897S	6902	TVP902S
6872	TVP872S	6876	TVP876S	6890	TVP890S	6898	TVP898S	6904	TVP904S
6873	TVP873S	6881	TVP881S	6891	TVP891S				

6906-7000   Leyland Fleetline FE30AGR   MCW   H43/33F   1978-79

6906	WDA906T	6925	WDA925T	6947	WDA947T	6965	WDA965T	6982	WDA982T
6907	WDA907T	6926	WDA926T	6949	WDA949T	6966	WDA966T	6983	WDA983T
6909	WDA909T	6928	WDA928T	6950	WDA950T	6967	WDA967T	6984	WDA984T
6911	WDA911T	6930	WDA930T	6951	WDA951T	6968	WDA968T	6985	WDA985T
6912	WDA912T	6931	WDA931T	6952	WDA952T	6969	WDA969T	6986	WDA986T
6913	WDA913T	6932	WDA932T	6954	WDA954T	6970	WDA970T	6987	WDA987T
6915	WDA915T	6933	WDA933T	6955	WDA955T	6971	WDA971T	6988	WDA988T
6916	WDA916T	6934	WDA934T	6957	WDA957T	6972	WDA972T	6989	WDA989T
6918	WDA918T	6935	WDA935T	6958	WDA958T	6973	WDA973T	6991	WDA991T
6919	WDA919T	147	WDA940T	6960	WDA960T	6976	WDA976T	6996	WDA996T
6920	WDA920T	6941	WDA941T	6961	WDA961T	6977	WDA977T	6998	WDA998T
6922	WDA922T	6942	WDA942T	6962	WDA962T	6978	WDA978T	6999	WDA999T
6923	WDA923T	6945	WDA945T	6964	WDA964T	142	WDA979T	7000	WDA700T
6924	WDA924T								

7007	BOM7V	MCW Metrobus DR104/4	MCW	H43/30F	1979

Previous Registrations:

245DOC	G778HOV	GIL2942	E355EVH	TIA5734	E357EVH
5010CD	G544JOG	HDZ8350	F920BVP	WLT702	G776HOV
787LOM	G669JOH	HDZ8352	F77XUH		

Liveries: Blue, silver and red; National Express Rapide: K2-5CEN
The following vehicles carry the traditional liveries of those municipal fleets that formed West Midlands PTE: 2856/72, 3000/30/43/50/88, 3225 - Birmingham; 2867 - Coventry; 2888 - Walsall; 2989 - Wolverhampton and 3033 - West Bromwich.
On Order - 100 Volvo B10L buses with Wright Liberator bodies and a number of Volvo B6LE midibuses.

WILLIAMSONS

H J Williamson, Unit 4, The Depot, Knockin Heath, Oswestry, SY10 8DZ

RGS93R	Leyland Leopard PSU3C/4R	Plaxton Supreme III	C53F	1977	Ex Hellyers Coaches, Fareham, 1989
TPJ282S	Bedford YLQ	Plaxton Supreme III	C45F	1978	Ex Parry, Ruyton XI Towns, 1993
GWC31T	Leyland Leopard PSU3E/4R	Plaxton Supreme IV	C53F	1979	Ex Frost, Leigh-on-Sea, 1988
BAW1T	Leyland Leopard PSU3E/4R	Duple Dominant	B63F	1979	Ex Elcock Reisen, Madeley, 1984
UBW788	Leyland PSU5C/4R	Plaxton Supreme IV	C53F	1981	Ex Hown, Barnoldswick, 1993
C167VRE	Ford Transit 190	PMT	B16F	1986	Ex PMT, 1987
D532HNW	Ford Transit 190	Carlyle	B16F	1986	Ex Star Line, Knutsford, 1989
E45JRF	Leyland Tiger TRCTL11/3RZ	Plaxton Paramount 3500 III	C53F	1988	Ex PMT, 1994
F59YBO	Leyland Tiger TRCTL11/3ARZM	Duple 320	C61F	1989	Ex Bebb, Llantwit Fardre, 1991
G381JBD	LAG G355Z	LAG Panoramic	C49FT	1989	
H761RNT	Van Hool T815H	Van Hool Alizée	C49FT	1991	
J327VAW	Dennis Dart 9.8SDL3004	Carlyle Dartline	B40F	1991	
J328VAW	Dennis Dart 9.8SDL3004	Carlyle Dartline	B40F	1991	
J297KFP	MAN 10-180	Caetano Algarve II	C53FT	1991	Ex Bradsell, Slough, 1995
K870ANT	Volvo B10M-60	Jonckheere Deauville P599	C53FT	1992	
K871ANT	Dennis Dart 9.8SDL3012	Marshall C37	B43F	1992	
K152DNT	EOS E180Z	EOS 200	C52FT	1993	
K153DNT	EOS E180Z	EOS 200	C52FT	1993	
M701HBC	Dennis Javelin 12SDA2131	Plaxton Première 320	C57F	1994	

Previous Registrations:
UBW788   SLJ388X

Livery: Green and yellow; yellow & blue (Shrewsbury Park & Ride) BAW1T, J327/8VAW, K871ANT.

Williamsons operate the Harlescott Park & Ride in Shrewsbury with a selection of vehicles in the blue and yellow dedicated livery. Negotiating Dogpole, one of the narrow streets in Shrewsbury is J327VAW, a Carlyle Dartline-bodied Dennis Dart. *Tony Wilson*

WORTHEN TRAVEL

D A Pye, Main Road, Worthen, Shrewsbury   SY5 9HW

MVS514	AEC Reliance 2MU3RA	Duple Britannia	C40F	1960	Ex preservation, 1993
MUR215L	Bedford YRT	Duple Dominant	C53F	1973	Ex Travelwell Cs, Llandudno Jct, 1996
JEA884N	Leyland Leopard PSU5/4R	Plaxton Elite III (1975)	C53F	1972	Ex Clews, Sutton, 1993
PJP933R	Bedford YMT	Duple Dominant II	C53F	1977	Ex Burton, Stanion, 1994
DBG259T	Leyland Leopard PSU5C/4R	Plaxton Supreme IV	C57F	1979	Ex Four Square Coaches, Brierley, 1996
FIL6676	DAF MB200DKTL600	Plaxton Supreme IV	C57F	1981	Ex Sattelite Coaches, Worcester, 1992
FO8933	DAF MB200DKTL600	Plaxton Paramount 3200	C57F	1983	Ex Morris Travel, Pencoed, 1991
OIJ6726	DAF SB2300DHS585	Plaxton Paramount 3500	C49FT	1984	Ex Hallam, Newthorpe, 1994
G852VAY	Dennis Javelin 12SDA1907	Duple 320	C57F	1989	Ex Owen's, Oswestry, 1994
G920WAY	DAF SB2305DHS585	Caetano Algarve II	C51F	1990	Ex Lewis, Greenoch, 1996
413MAB	Dennis Javelin 12SDA1907	Caetano Algarve	C57F	1989	Ex Owen's, Oswestry, 1994
H154OEP	Nissan Urvan	Deansgate	M10	1991	Ex Owen's, Oswestry, 1995

Previous Registrations:

413MAB	G853VAY		FO8933	THB430Y	JEA884N	DAR519K
FIL6676	PRM971X		MVS514	615HX	OIJ6726	A800LEL, XEL941, A264NFX

Livery: Two-tone Green

Representing the Worthen Travel fleet is Bova Futura AJF88A, which joined the fleet in 1989 from Silver Coach Lines of Edinburgh and recently left it as part of a continual modernisation. The vehicle is seen on National Express duplication work in Buckingham Palace Road, London. *Colin Lloyd*

ZAK'S

K P Fazakarley, 424 Beeches Road, Great Barr, Birmingham B42 2QW

PDN405P	Bedford YLQ	Plaxton Supreme III	C45F	1976	Ex Jolly Roger, New Earswick, 1994
NDT634X	Mercedes-Benz L207D	Whittaker	M12	1981	Ex Belgrave, Syston, 1986
HPN945	Volvo B10M-61	Van Hool Alizée	C50FT	1983	Ex HAD, Shotts, 1994
A60NPP	Mercedes-Benz L307D	Reeve Burgess	M12	1983	Ex Allen, Coventry, 1992
DSU109	Leyland Tiger TRCTL11/3R	Plaxton Paramount 3500	C49FT	1984	Ex Drayton vale, Ratby, 1993
B644DWA	Ford Transit 190	Ford	M8	1984	Ex private owner, 1992
B755ESE	Mercedes-Benz L608D	Reeve Burgess	C19F	1985	Ex Costelloe, Barnoldswick, 1989
B973NVT	Ford Transit 160	Ford	M8	1985	Ex private owner, 1992
C565TUT	Ford Transit 190	Dormobile	B16F	1986	Ex Delta, Stockton, 1994
C739HKK	Freight Rover Sherpa	Dormobile	B18F	1988	Ex Lionspeed, West Bromwich, 1994
C136HVW	Ford Transit 190	Lamocote	M12	1986	Ex Lamcote, Radcliffe, 1992
C448RUY	Renault-Dodge S56	Dormobile	B--FL	1986	Ex pricate owner, 1993
C345VVT	Ford Transit 190	Bedford Coachworks	M12	1986	Ex Neil Audley, 1992
D107TFT	Freight Rover Sherpa	Carlyle	B20F	1986	Ex RoadCar, 1994
D507NDA	Freight Rover Sherpa	Carlyle	B18F	1986	Ex Careline, Birmingham, 1992
D554JFK	Renault ---	Ford ??	M8	1987	Ex pricate owner, 1992
D549XRB	Ford Transit -	Ford	M12	1987	Ex private owner, 1991
E125RAX	Freight Rover Sherpa	Carlyle Citybus 2	B20F	1987	Ex Morris, Erskine, 1995, 1992
E104UEA	Ford Transit VE6	Ford	M14	1988	Ex private owner, 1992
F92JFJ	Renault Trafic	Devon Conversions	M8	1989	
G226EOA	Freight Rover Sherpa	Carlyle Citybus 2	B20F	1989	Ex Merry Hill Minibus, 1992
G373YUR	Iveco 315	Lorraine	C30F	1990	Ex Brents, Watford, 1992
G206YDL	Mercedes-Benz 811D	Pheonix	B31F	1990	Ex Solent Blue Line, 1995
H671ATN	Toyota Coaster HB31R	Caetano Optimo	C21F	1990	Ex Coussess & Castleton, Stainton, 1995

Previous Registrations:
DSU109   A779WHB   HPN945   MSU582Y

Livery: ??

Passing The Pallasades shopping complex near New Street rail station is Zak's recently delivered Leyland National GLJ682N.
Philip Lamb

Index to Vehicles

1FTG	Flight's	9995RU	Bakers	A674UOE	W M Buses	A734UOE	W M Buses
1KOV	Harry Shaw	A2FTG	Flight's	A675UOE	W M Buses	A735GFA	P M T
2FTG	Flight's	A3FTG	Flight's	A676UOE	W M Buses	A735UOE	W M Buses
3KOV	Harry Shaw	A4FTG	Flight's	A677UOE	W M Buses	A736GFA	P M T
55MDT	Travel de Courcey	A21FVT	Leon's	A678UOE	W M Buses	A736WVP	W M Buses
82HBC	Stevensons	A32GJT	NCB Motors	A679UOE	W M Buses	A737GFA	P M T
84COV	Harry Shaw	A33GJT	NCB Motors	A680UOE	W M Buses	A737WVP	W M Buses
86TN171	Travel de Courcey	A39SMA	Stevensons	A681UOE	W M Buses	A738GFA	P M T
86TN172	Travel de Courcey	A41SMA	Stevensons	A682UOE	W M Buses	A738WVP	W M Buses
86TS330	Travel de Courcey	A42SMA	Midland	A683UOE	W M Buses	A739GFA	P M T
88KK1114	Travel de Courcey	A60NPP	Zak's	A684UOE	W M Buses	A739WVP	W M Buses
88TN1224	Travel de Courcey	A101JJT	King Offa	A685UOE	W M Buses	A740GFA	P M T
123TKM	Midland	A101WVP	W M Buses	A686UOE	W M Buses	A740WVP	W M Buses
124YTW	Stevensons	A102WVP	W M Buses	A687UOE	W M Buses	A741GFA	P M T
222UPD	Birmingham Coach	A103WVP	W M Buses	A688UOE	W M Buses	A741WVP	W M Buses
245DOC	W M Buses	A105WVP	W M Buses	A689UOE	W M Buses	A742GFA	P M T
413MAB	Worthern Travel	A106WVP	W M Buses	A690UOE	W M Buses	A742WVP	W M Buses
422AKN	Stevensons	A107WVP	W M Buses	A690UOE	W M Buses	A743JRE	P M T
468KPX	Stevensons	A108WVP	W M Buses	A691UOE	W M Buses	A743WVP	W M Buses
479BOC	Stevensons	A109WVP	W M Buses	A692CHJ	Travel de Courcey	A744JRE	P M T
488BDN	Stevensons	A110WVP	W M Buses	A692UOE	W M Buses	A744WVP	W M Buses
507EXA	P M T	A111WVP	W M Buses	A693UOE	W M Buses	A745JRE	P M T
510DMY	Minsterley	A112WVP	W M Buses	A694UOE	W M Buses	A745WVP	W M Buses
565LON	Stevensons	A113WVP	W M Buses	A695UOE	W M Buses	A746JRE	P M T
614WEH	Stevensons	A114WVP	W M Buses	A696UOE	W M Buses	A746WVP	W M Buses
784RBF	Stevensons	A122BHL	Choice Travel	A697UOE	W M Buses	A747JRE	P M T
787LOM	W M Buses	A136SMA	P M T	A698UOE	W M Buses	A747WVP	W M Buses
803HOM	Stevensons	A137SMA	P M T	A699UOE	W M Buses	A748WVP	W M Buses
852YYC	Stevensons	A138SMA	P M T	A701HVT	Midland	A749WVP	W M Buses
904AXY	Stevensons	A139EPA	Midland	A701UOE	W M Buses	A750WVP	W M Buses
1223PL	Ludlows	A143SMA	P M T	A702HVT	Midland	A751WVP	W M Buses
1398NT	Elcock Reisen	A144SMA	P M T	A702UOE	W M Buses	A752WVP	W M Buses
1497RU	Bakers	A145SMA	P M T	A703HVT	Midland	A753WVP	W M Buses
1513RU	Bakers	A146UDM	P M T	A703UOE	W M Buses	A754WVP	W M Buses
1577NT	Elcock Reisen	A150UDM	Midland	A704HVT	Midland	A755WVP	W M Buses
1655VT	Blue Bus	A152UDM	Stevensons	A704UOE	W M Buses	A756WVP	W M Buses
1672VT	Blue Bus	A154UDM	Midland	A705HVT	Midland	A757WVP	W M Buses
1877NT	Minsterley	A155UDM	Midland	A705UOE	W M Buses	A758WVP	W M Buses
1879RU	Bakers	A156UDM	P M T	A706HVT	Midland	A759WVP	W M Buses
2335PL	Leon's	A157UDM	P M T	A706UOE	W M Buses	A760WVP	W M Buses
3093RU	Bakers	A158UDM	P M T	A707HVT	Midland	A761WVP	W M Buses
3102RU	Bakers	A159UDM	P M T	A707UOE	W M Buses	A762WVP	W M Buses
3275RU	Bakers	A160EPA	Midland	A708HVT	Midland	A763WVP	W M Buses
3353RU	Bakers	A160UDM	P M T	A708UOE	W M Buses	A764WVP	W M Buses
3408NT	Elcock Reisen	A161VDM	P M T	A709HVT	Midland	A765WVP	W M Buses
3471RU	Bakers	A162VDM	P M T	A709UOE	W M Buses	A766WVP	W M Buses
3563RU	Bakers	A163VDM	P M T	A710UOE	W M Buses	A767WVP	W M Buses
3566RU	Bakers	A164VDM	P M T	A712RCA	Leon's	A768WVP	W M Buses
3601RU	Bakers	A165VDM	P M T	A712UOE	W M Buses	A769WVP	W M Buses
4195PX	Boydon	A166VFM	P M T	A713UOE	W M Buses	A770WVP	W M Buses
4327PL	Leon's	A167VFM	P M T	A714UOE	W M Buses	A771WVP	W M Buses
5010CD	W M Buses	A168VFM	P M T	A715UOE	W M Buses	A772WVP	W M Buses
5038NT	Elcock Reisen	A169VFM	P M T	A716UOE	W M Buses	A845GEJ	M&J Travel
5621RU	Bakers	A170VFM	P M T	A717UOE	W M Buses	A891EBC	Boydon
5658RU	Bakers	A171VFM	P M T	A718UOE	W M Buses	A898KAH	Midland
5702PL	Leon's	A172VFM	Midland	A719UOE	W M Buses	AAK106T	Rest & Ride
5777RU	Bakers	A174NAC	Travel de Courcey	A720UOE	W M Buses	AAL272A	Birmingham Coach
5888EH	Leon's	A195KKF	Stevensons	A721UOE	W M Buses	AAL303A	Stevensons
6280RU	Bakers	A215PEV	Midland	A722UOE	W M Buses	AAL345A	Birmingham Coach
6577RU	Bakers	A386NNK	N C P	A723UOE	W M Buses	AAL404A	Stevensons
7025RU	Bakers	A512LPP	M&J Travel	A724UOE	W M Buses	AAL453A	Birmingham Coach
7092RU	Bakers	A529DNR	Warringtons	A725UOE	W M Buses	AAX562A	Stevensons
8150RU	Bakers	A531CUX	Jones	A726UOE	W M Buses	AAX590A	Midland
8399RU	Bakers	A573BRD	Serveverse	A727UOE	W M Buses	ABW210L	Ludlows
8636PL	Leon's	A615KRT	Minsterley	A728UOE	W M Buses	ADF871T	Boydon
8797PL	Ludlows	A653ANT	Shropshire Ed	A729UOE	W M Buses	AEF765A	Glenstuart
8830RU	Bakers	A668UOE	W M Buses	A730UOE	W M Buses	AFA729S	Procters
9346PL	Leon's	A669UOE	W M Buses	A731UOE	W M Buses	AHU515V	P M T
9423RU	Bakers	A670UOE	W M Buses	A732UOE	W M Buses	AHW203V	P M T
9530RU	Bakers	A671UOE	W M Buses	A733GFA	P M T	AHW206V	Stevensons
9595RU	Bakers	A672UOE	W M Buses	A733UOE	W M Buses	AHW207V	Stevensons
9685VT	Blue Bus	A673UOE	W M Buses	A734GFA	P M T	AKU160T	Birmingham Coach

116   *The North & West Midlands Bus Handbook*

Reg	Operator	Reg	Operator	Reg	Operator	Reg	Operator
AKU165T	Serveverse	B783AOC	W M Buses	B858AOP	W M Buses	BOK32V	W M Buses
AKY612T	Birmingham Coach	B784AOC	W M Buses	B859AOP	W M Buses	BOK33V	W M Buses
ANA8T	Knotty	B785AOC	W M Buses	B860AOP	W M Buses	BOK34V	W M Buses
AOL9T	W M Buses	B786AOC	W M Buses	B861DOM	W M Buses	BOK35V	W M Buses
AOL10T	W M Buses	B787AOC	W M Buses	B862DOM	W M Buses	BOK36V	W M Buses
AOL11T	W M Buses	B788AOC	W M Buses	B863DOM	W M Buses	BOK37V	W M Buses
AOL12T	W M Buses	B789AOC	W M Buses	B864DOM	W M Buses	BOK38V	W M Buses
AOL13T	W M Buses	B790AOC	W M Buses	B865DOM	W M Buses	BOK39V	W M Buses
AOL14T	W M Buses	B791AOC	W M Buses	B866DOM	W M Buses	BOK40V	W M Buses
AOL15T	W M Buses	B792AOC	W M Buses	B867DOM	W M Buses	BOK41V	W M Buses
AOL17T	W M Buses	B793AOC	W M Buses	B868DOM	W M Buses	BOK42V	W M Buses
APM106T	Knotty	B794AOC	W M Buses	B869DOM	W M Buses	BOK43V	W M Buses
ATH4V	Elcock Reisen	B795AOC	W M Buses	B870DOM	W M Buses	BOK44V	W M Buses
AUP650L	Boydon	B796AOC	W M Buses	B871DOM	W M Buses	BOK45V	W M Buses
AUT842Y	Butters	B797AOP	W M Buses	B872DOM	W M Buses	BOK46V	W M Buses
AYJ106T	Birmingham Coach	B798AOP	W M Buses	B873DOM	W M Buses	BOK47V	W M Buses
AYR309T	Chase	B799AOP	W M Buses	B874DOM	W M Buses	BOK48V	W M Buses
AYR317T	Chase	B800AOP	W M Buses	B875DOM	W M Buses	BOK49V	W M Buses
AYR324T	Falcon Travel	B801AOP	W M Buses	B875EOM	Midland	BOK50V	W M Buses
AYR330T	Chase	B802AOP	W M Buses	B876DOM	W M Buses	BOK51V	W M Buses
AYR339T	Chase	B803AOP	W M Buses	B877DOM	W M Buses	BOK52V	W M Buses
AYR343T	Chase	B804AOP	W M Buses	B878DOM	W M Buses	BOK53V	W M Buses
B53AOC	W M Buses	B805AOP	W M Buses	B879DOM	W M Buses	BOK54V	W M Buses
B54AOC	W M Buses	B806AOP	W M Buses	B880DOM	W M Buses	BOK55V	W M Buses
B102KPF	Midland	B807AOP	W M Buses	B881DOM	W M Buses	BOK56V	W M Buses
B103KPF	Midland	B808AOP	W M Buses	B882DOM	W M Buses	BOK57V	W M Buses
B104KPF	Midland	B809AOP	W M Buses	B882HSX	Stevensons	BOK58V	W M Buses
B105KPF	Midland	B810AOP	W M Buses	B883DOM	W M Buses	BOK59V	W M Buses
B108KPF	Midland	B811AOP	W M Buses	B884DOM	W M Buses	BOK60V	W M Buses
B109KPF	Midland	B812AOP	W M Buses	B885DOM	W M Buses	BOK61V	W M Buses
B117OBF	P M T	B813AOP	W M Buses	B886DOM	W M Buses	BOK62T	Ludlows
B145ALG	Stevensons	B814AOP	W M Buses	B911NBF	Midland	BOK62V	W M Buses
B147ALG	Stevensons	B815AOP	W M Buses	B912NBF	Midland	BOK63V	W M Buses
B148ALG	Stevensons	B816AOP	W M Buses	B913NBF	Midland	BOK64V	W M Buses
B149ALG	Stevensons	B817AOP	W M Buses	B973NVT	Zak's	BOK65V	W M Buses
B150ALG	Stevensons	B818AOP	W M Buses	B977HNT	NCB Motors	BOK66V	W M Buses
B181BLG	P M T	B819AOP	W M Buses	BAL607T	Birmingham Coach	BOK67V	W M Buses
B182BLG	P M T	B820AOP	W M Buses	BAL608T	Falcon Travel	BOK68V	Stevensons
B188BLG	P M T	B821AOP	W M Buses	BAL609T	Birmingham Coach	BOK69V	W M Buses
B192PFA	Leon's	B822AOP	W M Buses	BAW1T	Williamson	BOK70V	W M Buses
B195BLG	P M T	B823AOP	W M Buses	BGK314S	Britannia	BOK71V	W M Buses
B197DTU	Midland	B824AOP	W M Buses	BGR683W	Minsterley	BOK72V	Stevensons
B198DTU	Midland	B825AOP	W M Buses	BGR684W	Minsterley	BOK73V	W M Buses
B199DTU	P M T	B826AOP	W M Buses	BGY589T	Claribels	BOK74V	W M Buses
B200DTU	P M T	B827AOP	W M Buses	BHK565X	Sandwell Travel	BOK75V	Stevensons
B201DTU	P M T	B828AOP	W M Buses	BJT322T	Stevensons	BOK76V	W M Buses
B202DTU	P M T	B829AOP	W M Buses	BLJ720Y	Jones	BOK77V	W M Buses
B203DTU	Midland	B830AOP	W M Buses	BMA523W	Midland	BOK78V	W M Buses
B204DTU	Midland	B831AOP	W M Buses	BOK1V	W M Buses	BOK79V	W M Buses
B216GUX	Owen's	B832AOP	W M Buses	BOK2V	W M Buses	BOK80V	W M Buses
B232AFV	P M T	B833AOP	W M Buses	BOK3V	W M Buses	BOK81V	W M Buses
B250HUX	Jones	B834AOP	W M Buses	BOK5V	W M Buses	BOK82V	W M Buses
B341BBV	Shropshire Ed	B835AOP	W M Buses	BOK6V	W M Buses	BOK83V	W M Buses
B413NJF	Handybus	B836AOP	W M Buses	BOK7V	W M Buses	BOK84V	W M Buses
B414NJF	Handybus	B837AOP	W M Buses	BOK8V	W M Buses	BOK85V	W M Buses
B422NJF	Handybus	B838AOP	W M Buses	BOK9V	W M Buses	BOK86V	W M Buses
B516OEH	Midland	B839AOP	W M Buses	BOK10V	W M Buses	BOK87V	W M Buses
B604OEH	Midland	B840AOP	W M Buses	BOK11V	W M Buses	BOK88V	W M Buses
B605OEH	Midland	B841AOP	W M Buses	BOK12V	W M Buses	BOK89V	W M Buses
B606OEH	Midland	B841WYH	Sandwell Travel	BOK13V	W M Buses	BOK90V	W M Buses
B607OEH	Midland	B842AOP	W M Buses	BOK14V	W M Buses	BOM7V	W M Buses
B644DWA	Zak's	B843AOP	W M Buses	BOK15V	W M Buses	BPR102Y	Midland
B730YUD	Midland	B844AOP	W M Buses	BOK16V	W M Buses	BPR106Y	Midland
B732YUD	Midland	B845AOP	W M Buses	BOK17V	W M Buses	BPR107Y	Midland
B733YUD	Midland	B846AOP	W M Buses	BOK18V	W M Buses	BPT922S	Leon's
B734YUD	Midland	B847AOP	W M Buses	BOK19V	W M Buses	BRF689T	P M T
B755ESE	Zak's	B848AOP	W M Buses	BOK20V	W M Buses	BRF693T	P M T
B774AOC	W M Buses	B849AOP	W M Buses	BOK21V	W M Buses	BSF766S	Birmingham Coach
B775AOC	W M Buses	B850AOP	W M Buses	BOK22V	W M Buses	BSN878V	Stevensons
B776AOC	W M Buses	B851AOP	W M Buses	BOK23V	W M Buses	BTU565S	Longmynd
B777AOC	W M Buses	B852AOP	W M Buses	BOK25V	W M Buses	BVP763V	Midland
B778AOC	W M Buses	B853AOP	W M Buses	BOK26V	W M Buses	BVP764V	Midland
B779AOC	W M Buses	B854AOP	W M Buses	BOK27V	W M Buses	BVP765V	Midland
B780AOC	W M Buses	B855AOP	W M Buses	BOK28V	W M Buses	BVP767V	Midland
B781AOC	W M Buses	B856AOP	W M Buses	BOK30V	W M Buses	BVP802V	Travel de Courcey
B782AOC	W M Buses	B857AOP	W M Buses	BOK31V	W M Buses	BVP968V	Midland

The North & West Midlands Bus Handbook 117

BVP969V	Midland	C544TJF	Handybus	D102AFV	Serveverse	D307SDS	Metropilitan
BXI2410	Owen's	C565TUT	Zak's	D102CFA	Midland	D312SDS	Metropilitan
BYW357V	Chase	C573TUT	Moorland Buses	D104CFA	Midland	D314EFK	Patterson
BYW358V	Chase	C576TUT	Handybus	D105CFA	Midland	D314MHS	Metropilitan
BYW360V	Ludlows	C588UFA	Leon's	D107CFA	Midland	D315EFK	Patterson
BYW365V	Chase	C683LGE	P M T	D107OWG	Little Red Bus	D316EFK	Patterson
BYW366V	Chase	C706JMB	P M T	D107TFT	Zak's	D316MHS	Metropilitan
BYW369V	Chase	C711JMB	P M T	D109OWG	Banga Travel	D317EFK	Patterson
BYW382V	Chase	C725JJO	Metropilitan	D110CFA	Midland	D318EFK	Patterson
C29WBF	Handybus	C726JJO	Midland	D113TFT	Handybus	D319DEF	Midland
C37WBF	Midland	C739HKK	Zak's	D120PGA	P M T	D319EFK	Patterson
C43WBF	M&J Travel	C777KGB	Clowes	D121PGA	P M T	D322DEF	Midland
C55HOM	W M Buses	C802SDY	Stevensons	D122PGA	P M T	D401MHS	Stevensons
C56HOM	W M Buses	C805KBT	Sandwell Travel	D122TFT	Handybus	D401NNA	Midland
C57HOM	W M Buses	C817CBU	Lionspeed/Pete's	D122WCC	Handybus	D402NNA	Midland
C58HOM	W M Buses	C819CBU	Little Red Bus	D123PGA	P M T	D409NUH	Lionspeed/Pete's
C59HOM	W M Buses	C822CBU	Little Red Bus	D124WCC	Green Bus	D410NUH	Lionspeed/Pete's
C60HOM	W M Buses	C822SDY	Stevensons	D126OWG	Lionspeed/Pete's	D413GBF	Handybus
C61HOM	W M Buses	C823SDY	Stevensons	D133NUS	Stevensons	D414FEH	Lionspeed/Pete's
C62HOM	W M Buses	C824CBU	Lionspeed/Pete's	D135NUS	Stevensons	D416FEH	Lionspeed/Pete's
C63HOM	W M Buses	C832CBU	Lionspeed/Pete's	D135TFT	Handybus	D422NNA	Midland
C64HOM	W M Buses	C833CBU	Lionspeed/Pete's	D138LTA	Lionspeed/Pete's	D423GBF	Handybus
C65HOM	W M Buses	C834CBU	Lionspeed/Pete's	D141NUS	Stevensons	D423NNA	Happy Days
C66HOM	W M Buses	C887FON	W M Buses	D142RAK	Lionspeed/Pete's	D424POF	Bowens
C78WRE	Stevensons	C888FON	W M Buses	D148RAK	Lionspeed/Pete's	D429NNA	Midland
C85AUB	Midland	C889FON	W M Buses	D152BEH	P M T	D430NNA	Midland
C89AUB	Handybus	C890FON	W M Buses	D153BEH	P M T	D431TCA	Metropilitan
C108SFP	P M T	C891FON	W M Buses	D154BEH	P M T	D432TCA	Metropilitan
C120VBF	P M T	C892FON	W M Buses	D155BEH	P M T	D433TCA	Metropilitan
C121VRE	P M T	C893FON	W M Buses	D156BEH	P M T	D434NNA	Midland
C122VRE	P M T	C894FON	W M Buses	D156LTA	Copeland's	D438NNA	Midland
C123VRE	P M T	C895FON	W M Buses	D157BEH	P M T	D444NNA	Midland
C124LHS	P M T	C896FON	W M Buses	D157VRP	P M T	D448NNA	Midland
C124VRE	P M T	C897FON	W M Buses	D158BEH	P M T	D450NNA	Midland
C125VRE	P M T	C898FON	W M Buses	D158LTA	Lionspeed/Pete's	D451ERE	P M T
C126VRE	P M T	C899FON	W M Buses	D159BEH	P M T	D452ERE	P M T
C127VRE	P M T	C900FON	W M Buses	D159VRP	P M T	D453ERE	P M T
C128VRE	P M T	C901FON	W M Buses	D160VRP	P M T	D454ERE	P M T
C130VRE	P M T	C902FON	W M Buses	D161LTA	Lionspeed/Pete's	D455ERE	P M T
C131VRE	P M T	C903FON	W M Buses	D162LTA	P M T	D456ERE	P M T
C132VRE	P M T	C904FON	W M Buses	D164LTA	Copeland's	D457ERE	P M T
C133VRE	P M T	C905FON	W M Buses	D167NON	Moorland Buses	D458ERE	P M T
C134VRE	P M T	C906FON	W M Buses	D169RAK	Lionspeed/Pete's	D459ERE	P M T
C135VRE	P M T	C907FON	W M Buses	D175LTA	Lionspeed/Pete's	D4xxGBF	Handybus
C136HVW	Zak's	C908FON	W M Buses	D176LNA	Stevensons	D507NDA	Zak's
C136VRE	P M T	C909FON	W M Buses	D176VRP	P M T	D525NDA	Serveverse
C137VRE	P M T	C910FON	W M Buses	D179VRP	P M T	D526HNW	Happy Days
C138VRE	P M T	C914AWK	Travel de Courcey	D182BEH	P M T	D528NDA	Banga Travel
C139VRE	P M T	CAZ2747	Claribels	D183BEH	P M T	D532HNW	Williamson
C140VRE	P M T	CAZ2748	Claribels	D184VRP	P M T	D534FAE	Stevensons
C141SPB	Midland	CAZ2749	Claribels	D185BEH	P M T	D538FAE	Stevensons
C141VRE	P M T	CBK931W	Boydon	D185VRP	P M T	D548FAE	P M T
C142VRE	P M T	CBV785S	W M Buses	D186VRP	P M T	D549FAE	P M T
C143VRE	P M T	CDC168K	Horrocks	D188BEH	P M T	D549XRB	Zak's
C144VRE	P M T	CRE240T	Leon's	D190NON	Banga Travel	D550FAE	P M T
C145WRE	P M T	CRP310K	Knotty	D219OOJ	Midland	D553NOE	W M Buses
C146WRE	P M T	CVN347Y	Copeland's	D222NCS	Metropilitan	D554JFK	Zak's
C147WRE	P M T	CWA439T	Minsterley	D224NCS	Metropilitan	D554NOE	W M Buses
C148WRE	P M T	CYA181J	Knotty	D230POF	Bowens	D559HNW	Happy Days
C149WRE	P M T	CYH578V	Clowes	D231NCS	Metropilitan	D566NDA	W M Buses
C150WRE	P M T	D31SAO	Banga Travel	D233MKK	Sandwell Travel	D568NDA	W M Buses
C151WRE	P M T	D36KAX	Choice Travel	D234MKK	Sandwell Travel	D569NDA	W M Buses
C167VRE	Williamson	D39NDW	Lionspeed/Pete's	D234NCS	Metropilitan	D569RKW	M&J Travel
C230AEA	Leon's	D39TKA	Moorland Buses	D235MKK	Sandwell Travel	D575NDA	W M Buses
C252SPC	Stevensons	D60TLV	Happy Days	D236MKK	Sandwell Travel	D578NDA	W M Buses
C262SPC	Midland	D62NOF	Moorland Buses	D246VNL	Little Red Bus	D579EWS	Handybus
C314NNT	Boultons	D63MTG	Banga Travel	D255VNL	Little Red Bus	D579NDA	W M Buses
C337CHT	Patterson	D69NOF	Handybus	D256NCS	Metropilitan	D581NDA	W M Buses
C345VVT	Zak's	D69OKG	Banga Travel	D257OOJ	Lionspeed/Pete's	D582NDA	W M Buses
C377MAW	Longmynd	D88CFA	Midland	D257YBB	Little Red Bus	D584NDA	W M Buses
C448RUY	Zak's	D91CFA	Midland	D259OOJ	Lionspeed/Pete's	D585NDA	W M Buses
C457AHY	Patterson	D92CFA	Midland	D302MHS	Banga Travel	D586NDA	W M Buses
C468TAY	Moorland Buses	D95CFA	Midland	D302SDS	Metropilitan	D588NDA	W M Buses
C477EUA	P M T	D96CFA	Midland	D303MHS	Metropilitan	D589NDA	W M Buses
C503PSC	Banga Travel	D98CFA	Midland	D304SDS	Metropilitan	D591NDA	W M Buses
C505PSC	Banga Travel	D101UJC	Handybus	D305JJD	Patterson	D592NDA	W M Buses

The latest livery style for the Glenstuart Travel fleet is seen in this picture of Plaxton Supreme 16, AEF765A pictured while on private hire duties. This vehicle is the only full-size coach in the fleet which mainly consists of minibuses and Leyland Nationals. *Tim Weatherup.*

D593NDA	W M Buses	D632NOE	W M Buses	D707TWM	Little Red Bus	D917NDA	W M Buses
D594NDA	W M Buses	D633NOE	W M Buses	D710SKU	Clowes	D918NDA	W M Buses
D595NDA	W M Buses	D634NOE	W M Buses	D710TWM	Lionspeed/Pete's	D918NDB	Little Red Bus
D596NDA	W M Buses	D635NOE	W M Buses	D711TWM	Lionspeed/Pete's	D919NDA	W M Buses
D598VBV	Moorland Buses	D636NOE	W M Buses	D712TWM	Little Red Bus	D919NDB	Little Red Bus
D601NOE	W M Buses	D637NOE	W M Buses	D713TWM	Lionspeed/Pete's	D920NDA	W M Buses
D602NOE	W M Buses	D640NOE	W M Buses	D715TWM	Little Red Bus	D921NDA	W M Buses
D603NOE	W M Buses	D641NOE	W M Buses	D720TNT	Jones	D922NDA	W M Buses
D605NOE	W M Buses	D642NOE	W M Buses	D735JUB	Sandwell Travel	D923NDA	W M Buses
D607NOE	W M Buses	D643NOE	W M Buses	D770JUB	Clowes	D924NDA	W M Buses
D608NOE	W M Buses	D644NOE	W M Buses	D780FVT	Bakers	D925KWW	Banga Travel
D609NOE	W M Buses	D646NOE	W M Buses	D780RBU	Little Red Bus	D925NDA	W M Buses
D611NOE	W M Buses	D647NOE	W M Buses	D784SGB	Bowens	D925NDB	Little Red Bus
D612NOE	W M Buses	D648NOE	W M Buses	D785SGB	Bowens	D926KWW	Banga Travel
D613NOE	W M Buses	D649NOE	W M Buses	D786SGB	Bowens	D926NDA	W M Buses
D614NOE	W M Buses	D650NOE	W M Buses	D787SGB	Bowens	D927NDA	W M Buses
D615NOE	W M Buses	D669SEM	Lionspeed/Pete's	D788JUB	Moorland Buses	D928NDA	W M Buses
D616NOE	W M Buses	D676MHS	Stevensons	D810NWW	P M T	D928NDB	Little Red Bus
D617NOE	W M Buses	D678MHS	Stevensons	D821RYS	Lionspeed/Pete's	D929NDA	W M Buses
D618NOE	W M Buses	D680MHS	Stevensons	D822PUK	Lionspeed/Pete's	D929NDB	Little Red Bus
D619NOE	W M Buses	D682MHS	Stevensons	D828PUK	Lionspeed/Pete's	D930NDA	W M Buses
D620NOE	W M Buses	D683MHS	Stevensons	D858LND	Lionspeed/Pete's	D931NDA	W M Buses
D621NOE	W M Buses	D691FDH	Patterson	D866NVS	Chase	D932NDA	W M Buses
D622NOE	W M Buses	D692FDH	Patterson	D870MDB	Lionspeed/Pete's	D932ODA	W M Buses
D623NOE	W M Buses	D693FDH	Patterson	D871MDB	Little Red Bus	D933NDA	W M Buses
D624NOE	W M Buses	D694FDH	Patterson	D900MWR	Claribels	D934NDA	W M Buses
D625BCK	Horrocks	D695FDH	Patterson	D911NDA	W M Buses	D935NDA	W M Buses
D625NOE	W M Buses	D696FDH	Patterson	D912NBA	Serveverse	D935NDB	Lionspeed/Pete's
D626NOE	W M Buses	D697FDH	Patterson	D912NDA	W M Buses	D935NDB	Little Red Bus
D627BCK	Patterson	D698FDH	Patterson	D913NDA	W M Buses	D936NDA	W M Buses
D627NOE	W M Buses	D699FDH	Patterson	D914NDA	W M Buses	D936NDB	Little Red Bus
D628NOE	W M Buses	D700FDH	Patterson	D914NDB	Little Red Bus	D937NDA	W M Buses
D629NOE	W M Buses	D701GHT	Patterson	D915NDA	W M Buses	D937NDB	Little Red Bus
D631NOE	W M Buses	D705GHT	Patterson	D916NDA	W M Buses	D938NDA	W M Buses

The North & West Midlands Bus Handbook

PMT's intake of new buses continues with fleet-liveried Dennis Lance SDC865, N865CEH and one of the first with the final version of the new fleet name. *Cliff Beeton*

D939NDA	W M Buses	DDW429V	Chase	E39KRE	P M T	E200TVE	Merry Hill
D940NDA	W M Buses	DJF633T	Horrocks	E41JRF	P M T	E222WUX	Elcock Reisen
D941NDA	W M Buses	DJN25X	Midland	E44JRF	P M T	E225WWD	Chase
D942NDA	W M Buses	DNT527T	King Offa	E45JRF	Williamson	E231NFX	Midland
D943NDA	W M Buses	DOC21V	W M Buses	E52MTC	Patterson	E232NFX	Green Bus
D943NDB	Lionspeed/Pete's	DOC22V	W M Buses	E72KBF	Stevensons	E234NFX	Midland
D944BAB	Patterson	DOC25V	W M Buses	E76PEE	Patterson	E261WWD	Elcock Reisen
D944NDA	W M Buses	DOC26V	W M Buses	E79OUH	Glenstuart	E269BRG	Lionspeed/Pete's
D945NDA	W M Buses	DOC28V	W M Buses	E83OUH	Glenstuart	E274HRY	Bowens
D945OKK	M&J Travel	DOC29V	W M Buses	E87OUH	Lionspeed/Pete's	E276HRY	Bowens
D946NDA	W M Buses	DOC35V	W M Buses	E90WCM	Midland	E277BRG	Little Red Bus
D946NDB	Little Red Bus	DOC37V	W M Buses	E91WCM	Midland	E278BRG	Little Red Bus
D947NDA	W M Buses	DOC39V	W M Buses	E93MRF	Bassetts	E290OMG	W M Buses
D948NDA	W M Buses	DOC47V	W M Buses	E93WCM	Midland	E311OMG	Longmynd
D949NDA	W M Buses	DOC48V	W M Buses	E94WCM	Midland	E312HLN	Patterson
D950NDA	W M Buses	DOC49V	W M Buses	E95WCM	Midland	E316NSX	Lionspeed/Pete's
D951NDA	W M Buses	DOC50V	W M Buses	E96WCM	Midland	E325JVN	Midland
D952NDA	W M Buses	DOC51V	W M Buses	E97WCM	Midland	E342NFA	P M T
D953NDA	W M Buses	DOC52V	W M Buses	E98LBC	M&J Travel	E384XCA	P M T
D954NDA	W M Buses	DSJ307V	Ludlows	E98WCM	Midland	E402YNT	Boultons
D955NDA	W M Buses	DSU109	Zak's	E99WCM	Midland	E406YMR	Metropilitan
D956NDA	W M Buses	DUH76V	Green Bus	E104UEA	Zak's	E407YMR	Metropilitan
D957NDA	W M Buses	DUH77V	Green Bus	E106LVT	P M T	E410EPE	Banga Travel
D958NDA	W M Buses	DUH78V	Green Bus	E110JPL	Midland	E467VNT	Boultons
D959NDA	W M Buses	DVT167J	Knotty	E125RAX	Zak's	E470MVT	P M T
D960NDA	W M Buses	DXI1454	Butters	E149RNY	Green Bus	E471MVT	P M T
D972PJA	Little Red Bus	E25UNE	Midland	E151AJC	Midland	E478NSC	Stevensons
D974TKC	Lionspeed/Pete's	E26UNE	Midland	E160NEU	Patterson	E504KNV	Longmynd
D976PJA	Little Red Bus	E27UNE	Midland	E163TWO	Blue Bus	E511TOV	W M Buses
DAR132T	Birmingham Coach	E28UNE	Midland	E170OMD	Longmynd	E512TOV	W M Buses
DBG259T	Worthern Travel	E29UNE	Midland	E176UWF	Lionspeed/Pete's	E514TOV	Merry Hill
DCA526X	P M T	E30UNE	Midland	E185UWF	Lionspeed/Pete's	E526NEH	P M T
DDM22X	King Offa	E31UNE	Midland	E196UKG	Blue Bus	E527JRE	P M T
DDW65V	Green Bus	E32UNE	Midland	E197UKG	Blue Bus	E528JRE	P M T

120 *The North & West Midlands Bus Handbook*

E533UOK	Patterson	E826HBF	P M T	F30XOF	W M Buses	F78XOF	W M Buses
E536PRU	Boultons	E829AWA	Stevensons	F31COM	Bowens	F79XOF	W M Buses
E542MRE	Bassetts	E831ETY	P M T	F31XOF	W M Buses	F80XOF	W M Buses
E564YBU	Stevensons	E834EVS	Chase	F32XOF	W M Buses	F81XOF	W M Buses
E590LEH	Leon's	E880YNT	Owen's	F33ENF	Midland	F82XOF	W M Buses
E599UHS	Bowens	E915NAC	W M Buses	F33XOF	W M Buses	F83XOF	W M Buses
E611LFV	Midland	E916NAC	W M Buses	F34ENF	Midland	F84XOF	W M Buses
E624FLD	N C P	E917NAC	W M Buses	F34XOF	W M Buses	F85XOF	W M Buses
E626FLD	N C P	E918NAC	W M Buses	F35ENF	Midland	F86XOF	W M Buses
E627FLD	N C P	E961YUX	Elcock Reisen	F35XOF	W M Buses	F87XOF	W M Buses
E635DCK	Little Red Bus	E969SOF	Moorland Buses	F36ENF	Midland	F88CWG	P M T
E637KCX	Claribels	E969SVP	Blue Bus	F36XOF	W M Buses	F88XOF	W M Buses
E638DCK	Little Red Bus	E975VUK	W M Buses	F37XOF	W M Buses	F89XOF	W M Buses
E639DCK	Little Red Bus	E976VUK	W M Buses	F38CWY	Rest & Ride	F90XOF	W M Buses
E645DCK	Little Red Bus	E977VUK	W M Buses	F38XOF	W M Buses	F91XOF	W M Buses
E650JWP	Boultons	E978VUK	W M Buses	F39ENF	Midland	F92JFJ	Zak's
E651RVP	W M Buses	E979VUK	W M Buses	F39HOD	Midland	F92XOF	W M Buses
E652RVP	W M Buses	E980VUK	W M Buses	F39XOF	W M Buses	F93XOF	W M Buses
E653RVP	W M Buses	E981VUK	W M Buses	F40ENF	Midland	F94XOF	W M Buses
E654SOL	W M Buses	E982VUK	W M Buses	F40XOF	W M Buses	F95CWG	P M T
E655RVP	W M Buses	E983VUK	W M Buses	F41XOF	W M Buses	F95XOF	W M Buses
E656RVP	W M Buses	E984VUK	W M Buses	F42XOF	W M Buses	F96PRE	Stevensons
E657RVP	W M Buses	E985VUK	W M Buses	F43XOF	W M Buses	F96XOF	W M Buses
E658RVP	W M Buses	E986VUK	W M Buses	F44XOF	W M Buses	F97PRE	Stevensons
E659RVP	W M Buses	E987VUK	W M Buses	F44XVP	Midland	F97XOF	W M Buses
E660RVP	W M Buses	E988VUK	W M Buses	F45XOF	W M Buses	F98XOF	W M Buses
E661RVP	W M Buses	E989SJA	Little Red Bus	F46XOF	W M Buses	F99XOF	W M Buses
E662RVP	W M Buses	E989VUK	W M Buses	F47XOF	W M Buses	F100UEH	P M T
E663RVP	W M Buses	E990NMK	Stevensons	F48XOF	W M Buses	F100XOF	W M Buses
E664RVP	W M Buses	E991SJA	Little Red Bus	F49XOF	W M Buses	F101XOF	W M Buses
E665RVP	W M Buses	E992NMK	Stevensons	F50XOF	W M Buses	F102XOF	W M Buses
E666YAW	Elcock Reisen	E993NMK	Stevensons	F50XOF	W M Buses	F103XOF	W M Buses
E673JNR	Bowens	EAA830W	Butters	F51ENF	Midland	F106CWG	P M T
E700HLB	Blue Bus	EBF806Y	Bassetts	F51XOF	W M Buses	F107CWG	P M T
E727HBF	Lionspeed/Pete's	EBM449T	Boydon	F52ENF	Midland	F108CWG	P M T
E728HBF	Lionspeed/Pete's	EBZ5229	Happy Days	F52XOF	W M Buses	F109CWG	P M T
E746JAY	Boultons	EDF274T	Jones	F53XOF	W M Buses	F110CWG	P M T
E749NSE	Boultons	EDJ242J	Knotty	F54XOF	W M Buses	F110SRF	Stevensons
E752XHL	Leon's	EEH902Y	Stevensons	F55XOF	W M Buses	F148USX	Midland
E760HBF	P M T	EEH903Y	Midland	F56XOF	W M Buses	F155DKU	Stevensons
E761HBF	P M T	EEH904Y	Stevensons	F57XOF	W M Buses	F161DET	Birmingham Coach
E762HBF	P M T	EEH905Y	Midland	F58XOF	W M Buses	F166DNT	P M T
E764HBF	P M T	EEH906Y	Midland	F59XOF	W M Buses	F170DET	Stevensons
E765HBF	P M T	EEH907Y	Midland	F59YBO	Williamson	F181YDA	Stevensons
E766HBF	P M T	EEH909Y	Midland	F61PRE	Stevensons	F185PRE	Stevensons
E767HBF	P M T	EEH910Y	Midland	F61XOF	W M Buses	F186PRE	Stevensons
E768HBF	P M T	EGB78T	Chase	F62XOF	W M Buses	F187REH	Stevensons
E769HBF	P M T	EGB81T	Choice Travel	F63XOF	W M Buses	F188REH	Stevensons
E783SJA	Lionspeed/Pete's	EGB90T	Birmingham Coach	F64XOF	W M Buses	F189RRF	Stevensons
E785SJA	Lionspeed/Pete's	EGB91T	Birmingham Coach	F65XOF	W M Buses	F190RRF	Stevensons
E791CCA	P M T	EGB93T	Glenstuart	F66XOF	W M Buses	F191SRF	Stevensons
E796SJA	Little Red Bus	EGB94T	Chase	F67DDA	W M Buses	F192VFA	Stevensons
E801HBF	P M T	EIL1607	Elcock Reisen	F67XOF	W M Buses	F203RAE	Patterson
E802HBF	P M T	EIL2247	Elcock Reisen	F68DDA	W M Buses	F210PNR	Longmynd
E803HBF	P M T	EIL829	Elcock Reisen	F68XOF	W M Buses	F217OFB	P M T
E804HBF	P M T	EJO490V	P M T	F69DDA	W M Buses	F233BAX	Lionspeed/Pete's
E805HBF	P M T	EJO491V	P M T	F69XOF	W M Buses	F249DKG	Green Bus
E806HBF	P M T	EJO492V	P M T	F70DDA	W M Buses	F258BHF	Owen's
E807HBF	P M T	EPD532V	Serveverse	F70XOF	W M Buses	F258GWJ	Stevensons
E808HBF	P M T	ERF24Y	P M T	F71DDA	W M Buses	F275CEY	Midland
E809HBF	P M T	ERP559T	Choice Travel	F71XOF	W M Buses	F276CEY	Midland
E810HBF	P M T	ESU241	Travel de Courcey	F72DDA	W M Buses	F278HOD	Midland
E811HBF	P M T	ETA978Y	Shropshire Ed	F72XOF	W M Buses	F300XOF	W M Buses
E812HBF	P M T	EVT690Y	Procters	F73DDA	W M Buses	F301XOF	W M Buses
E813HBF	P M T	EWY78Y	P M T	F73XOF	W M Buses	F302XOF	W M Buses
E814HBF	P M T	EWY79Y	P M T	F74DDA	W M Buses	F303XOF	W M Buses
E815HBF	P M T	EX9779	Little Red Bus	F74XOF	W M Buses	F304XOF	W M Buses
E816HBF	P M T	F22XOF	W M Buses	F75AKB	Metropolitan	F305XOF	W M Buses
E817HBF	P M T	F23XOF	W M Buses	F75DDA	W M Buses	F306XOF	W M Buses
E818HBF	P M T	F24XOF	W M Buses	F75XOF	W M Buses	F307XOF	W M Buses
E819HBF	P M T	F25XOF	W M Buses	F76DDA	W M Buses	F308XOF	W M Buses
E820HBF	P M T	F26XOF	W M Buses	F76XOF	W M Buses	F309XOF	W M Buses
E821HBF	P M T	F27XOF	W M Buses	F77DDA	W M Buses	F310REH	P M T
E822HBF	P M T	F28XOF	W M Buses	F77ERJ	Stevensons	F310XOF	W M Buses
E824HBF	P M T	F29XOF	W M Buses	F77XOF	W M Buses	F311REH	P M T
E825HBF	P M T	F30COM	Bowens	F78DDA	W M Buses	F311XOF	W M Buses

The North & West Midlands Bus Handbook

F312REH	P M T	F675YOG	W M Buses	G79EOG	W M Buses	G123EOG	W M Buses
F312XOF	W M Buses	F676YOG	W M Buses	G80EOG	W M Buses	G123FJW	W M Buses
F313REH	P M T	F677YOG	W M Buses	G81EOG	W M Buses	G124EOG	W M Buses
F313XOF	W M Buses	F678YOG	W M Buses	G82EOG	W M Buses	G124FJW	W M Buses
F314XOF	W M Buses	F679YOG	W M Buses	G83EOG	W M Buses	G125EOG	W M Buses
F315REH	P M T	F680YOG	W M Buses	G84EOG	W M Buses	G126EOG	W M Buses
F315XOF	W M Buses	F681YOG	W M Buses	G85EOG	W M Buses	G126TJA	Midland
F316REH	P M T	F682YOG	W M Buses	G86EOG	W M Buses	G127EOG	W M Buses
F316XOF	W M Buses	F683YOG	W M Buses	G87EOG	W M Buses	G127TJA	Midland
F317REH	P M T	F684YOG	W M Buses	G88EOG	W M Buses	G128EOG	W M Buses
F317XOF	W M Buses	F685YOG	W M Buses	G89EOG	W M Buses	G128TJA	Midland
F318XOF	W M Buses	F696ONR	Bowens	G90EOG	W M Buses	G129EOG	W M Buses
F319XOF	W M Buses	F700LCA	Midland	G91EOG	W M Buses	G130EOG	W M Buses
F320XOF	W M Buses	F703LCA	Midland	G92EOG	W M Buses	G131EOG	W M Buses
F321XOF	W M Buses	F705LCA	Midland	G92RGG	Jones	G132EOG	W M Buses
F326PPO	Stevensons	F713OFH	P M T	G93EOG	W M Buses	G133EOG	W M Buses
F335RWK	W M Buses	F792DWT	Stevensons	G93RGG	Jones	G134EOG	W M Buses
F336RWK	W M Buses	F811RJF	Travel de Courcey	G94EOG	W M Buses	G135EOG	W M Buses
F337RWK	W M Buses	F822GDT	Stevensons	G95EOG	W M Buses	G136EOG	W M Buses
F338RWK	W M Buses	F848RHY	Patterson	G96EOG	W M Buses	G136YRY	P M T
F361YTJ	P M T	F851RHY	Patterson	G97EOG	W M Buses	G137EOG	W M Buses
F362YTJ	P M T	F877RFP	Bassetts	G98EOG	W M Buses	G138EOG	W M Buses
F363YTJ	P M T	F877XOE	Merry Hill	G98VMM	Stevensons	G139EOG	W M Buses
F364YTJ	P M T	F878RFP	Bassetts	G99EOG	W M Buses	G140EOG	W M Buses
F372KBW	W M Buses	F878XOE	Merry Hill	G100EOG	W M Buses	G141EOG	W M Buses
F374DUX	NCB Motors	F879RFP	Bassetts	G101EOG	W M Buses	G141GOL	Stevensons
F452YHF	P M T	F879XOE	Merry Hill	G101EVT	P M T	G141LRM	Patterson
F472RBF	P M T	F881XOE	Merry Hill	G102EOG	W M Buses	G142EOG	W M Buses
F472RPG	Claribels	F882XOE	Merry Hill	G103EOG	W M Buses	G142LRM	Handybus
F473RBF	P M T	F883XOE	Merry Hill	G104EOG	W M Buses	G143EOG	W M Buses
F475VEH	P M T	F884XOE	Merry Hill	G104FJW	W M Buses	G144EOG	W M Buses
F482WFX	Elcock Reisen	F885XOE	Merry Hill	G105EOG	W M Buses	G145EOG	W M Buses
F483EJC	Midland	F886XOE	Merry Hill	G105FJW	W M Buses	G146EOG	W M Buses
F484EJC	Midland	F887XOE	Merry Hill	G106AVX	Owen's	G147EOG	W M Buses
F485EJC	Midland	F888XOE	Merry Hill	G106EOG	W M Buses	G147LRM	Patterson
F486EJC	Midland	F889XOE	Lionspeed/Pete's	G106FJW	W M Buses	G148EOG	W M Buses
F513RTL	Bassetts	F895XOE	Lionspeed/Pete's	G107EOG	W M Buses	G149EOG	W M Buses
F527BUX	Boultons	F907PFH	Stevensons	G108EOG	W M Buses	G149GOL	Choice Travel
F531UVT	P M T	F907UPR	Jones	G108FJW	W M Buses	G150EOG	W M Buses
F574YSC	Patterson	F929GGE	Patterson	G109EOG	W M Buses	G150GOL	Midland
F584BAW	Longmynd	F956XCK	Stevensons	G109FJW	W M Buses	G151EOG	W M Buses
F600EAW	Boultons	F968XWM	Patterson	G109YRE	Stevensons	G152EOG	W M Buses
F601EHA	Midland	F969HGE	Longmynd	G110EOG	W M Buses	G153EOG	W M Buses
F602EHA	Midland	F971HGE	NCB Motors	G110FJW	W M Buses	G154EOG	W M Buses
F603EHA	Midland	F985EDS	Stevensons	G111EOG	W M Buses	G155EOG	W M Buses
F604EHA	Midland	F990XOE	W M Buses	G111FJW	W M Buses	G155XJF	Claribels
F605EHA	Midland	F991XOE	W M Buses	G111TND	Midland	G156EOG	W M Buses
F606EHA	Midland	F992XOE	W M Buses	G112EOG	W M Buses	G157EOG	W M Buses
F607EHA	Midland	F993XOE	W M Buses	G112FJW	W M Buses	G158EOG	W M Buses
F608EHA	Midland	F994XOE	W M Buses	G113EOG	W M Buses	G159EOG	W M Buses
F608WBV	P M T	F995XOE	W M Buses	G113FJW	W M Buses	G160EOG	W M Buses
F609EHA	Midland	F996XOE	W M Buses	G114EOG	W M Buses	G160XJF	Longmynd
F610EHA	Midland	F997XOE	W M Buses	G114FJW	W M Buses	G161EOG	W M Buses
F611EHA	Midland	F998XOE	W M Buses	G114TND	Midland	G162EOG	W M Buses
F612EHA	Midland	F999PLA	Minsterley	G115EOG	W M Buses	G163EOG	W M Buses
F613EHA	Midland	F999XOE	W M Buses	G115FJW	W M Buses	G164EOG	W M Buses
F614EHA	Midland	FAZ3195	Stevensons	G115TND	Midland	G164YRE	Stevensons
F615EHA	Midland	FAZ5279	Stevensons	G116EOG	W M Buses	G165EOG	W M Buses
F616EHA	Midland	FBC473T	Knotty	G116FJW	W M Buses	G165YRE	Stevensons
F619EHA	Midland	FFR169S	Happy Days	G117EOG	W M Buses	G166EOG	W M Buses
F620EHA	Midland	FIL6676	Worthern Travel	G117FJW	W M Buses	G166YRE	Stevensons
F622EHA	Midland	FIL8602	Travel de Courcey	G117OGA	Leon's	G167EOG	W M Buses
F623EHA	Midland	FO8933	Worthern Travel	G117TND	Midland	G167YRE	Stevensons
F624EHA	Midland	FSV428	Chase	G118EOG	W M Buses	G168EOG	W M Buses
F625EHA	Midland	FTG5	Flight's	G118FJW	W M Buses	G168YRE	Stevensons
F626EHA	Midland	FTG9	Flight's	G119EOG	W M Buses	G169EOG	W M Buses
F660EBU	P M T	FWA499V	Happy Days	G119FJW	W M Buses	G169YRE	Stevensons
F666YOG	W M Buses	FXI8653	P M T	G120EOG	W M Buses	G170EOG	W M Buses
F667YOG	W M Buses	FYX812W	Claribels	G120FJW	W M Buses	G170YRE	Stevensons
F668YOG	W M Buses	FYX813W	Claribels	G121EOG	W M Buses	G171EOG	W M Buses
F669YOG	W M Buses	FYX817W	W M Buses	G121FJW	W M Buses	G171YRE	Stevensons
F670YOG	W M Buses	FYX818W	W M Buses	G121TJA	Midland	G172EOG	W M Buses
F671YOG	W M Buses	FYX823W	Claribels	G122DRF	Stevensons	G172YRE	Stevensons
F672YOG	W M Buses	G21YVT	Stevensons	G122EOG	W M Buses	G173EOG	W M Buses
F673YOG	W M Buses	G25YVT	Stevensons	G122FJW	W M Buses	G173YRE	Stevensons
F674YOG	W M Buses	G30POD	Longmynd	G122TJA	Midland	G174EOG	W M Buses

122 *The North & West Midlands Bus Handbook*

G175EOG	W M Buses	G241EOG	W M Buses	G306DPA	Midland	G708HOP	W M Buses
G176EOG	W M Buses	G242EOG	W M Buses	G306EOG	W M Buses	G709HOP	W M Buses
G177EOG	W M Buses	G243EOG	W M Buses	G307DPA	Midland	G710HOP	W M Buses
G178EOG	W M Buses	G244EOG	W M Buses	G307EOG	W M Buses	G711HOP	W M Buses
G179EOG	W M Buses	G245EOG	W M Buses	G308DPA	Midland	G712HOP	W M Buses
G180EOG	W M Buses	G246EOG	W M Buses	G308EOG	W M Buses	G713HOP	W M Buses
G181EOG	W M Buses	G247EOG	W M Buses	G309DPA	Midland	G714HOP	W M Buses
G182EOG	W M Buses	G248EOG	W M Buses	G309EOG	W M Buses	G715HOP	M&J Buses
G183DRF	Stevensons	G249EOG	W M Buses	G310DPA	Midland	G716HOP	W M Buses
G183EOG	W M Buses	G250EOG	W M Buses	G310EOG	W M Buses	G717HOP	W M Buses
G184DRF	Stevensons	G251EOG	W M Buses	G311EOG	W M Buses	G727RGA	Stevensons
G184EOG	W M Buses	G252EOG	W M Buses	G312EOG	W M Buses	G737NNS	Metropilitan
G185EOG	W M Buses	G253EOG	W M Buses	G313EOG	W M Buses	G738VKK	Minsterley
G186EOG	W M Buses	G254EOG	W M Buses	G314EOG	W M Buses	G753XRE	P M T
G187EOG	W M Buses	G255EOG	W M Buses	G315EOG	W M Buses	G754XRE	P M T
G188EOG	W M Buses	G256EOG	W M Buses	G316EOG	W M Buses	G755XRE	P M T
G189EOG	W M Buses	G257EOG	W M Buses	G318YVT	P M T	G756XRE	P M T
G190EOG	W M Buses	G258EOG	W M Buses	G327PHA	Midland	G757XRE	P M T
G191EOG	W M Buses	G259EOG	W M Buses	G328PHA	Midland	G758XRE	P M T
G192EOG	W M Buses	G260EOG	W M Buses	G330XRE	P M T	G759XRE	P M T
G193EOG	W M Buses	G261EOG	W M Buses	G331XRE	P M T	G760XRE	P M T
G194EOG	W M Buses	G262EOG	W M Buses	G332XRE	P M T	G761XRE	P M T
G195EOG	W M Buses	G263EOG	W M Buses	G333JUX	Elcock Reisen	G762XRE	P M T
G196EOG	W M Buses	G263GKG	Merry Hill	G333XRE	P M T	G785PWL	Stevensons
G197EOG	W M Buses	G264EOG	W M Buses	G334XRE	P M T	G801THA	Midland
G198EOG	W M Buses	G264GKG	Merry Hill	G335XRE	P M T	G802THA	Midland
G199EOG	W M Buses	G265EOG	W M Buses	G336XRE	P M T	G805AAD	P M T
G200EOG	W M Buses	G265GKG	Merry Hill	G337XRE	P M T	G807FJX	Stevensons
G201EOG	W M Buses	G266EOG	W M Buses	G338XRE	P M T	G838LWR	Boultons
G202EOG	W M Buses	G267EOG	W M Buses	G339XRE	P M T	G839GNV	Longmynd
G203EOG	W M Buses	G267GKG	Merry Hill	G340XRE	P M T	G852VAY	Worthern Travel
G204EOG	W M Buses	G268EOG	W M Buses	G341XRE	P M T	G864MYA	M&J Travel
G205EOG	W M Buses	G269EOG	W M Buses	G342CBF	P M T	G880ELJ	Chase
G206EOG	W M Buses	G270EOG	W M Buses	G343CBF	P M T	G897TGG	Midland
G206YDL	Zak's	G270GKG	Merry Hill	G344CBF	P M T	G900TJA	Midland
G207EOG	W M Buses	G271EOG	W M Buses	G345CBF	P M T	G901MNS	Stevensons
G208EOG	W M Buses	G271GKG	Merry Hill	G346CBF	P M T	G916LHA	Midland
G209EOG	W M Buses	G272EOG	W M Buses	G347ERF	P M T	G917LHA	Midland
G210EOG	W M Buses	G272GKG	Merry Hill	G348ERF	P M T	G918LHA	Midland
G211EOG	W M Buses	G273EOG	W M Buses	G349ERF	P M T	G919LHA	Midland
G212EOG	W M Buses	G273HBO	Merry Hill	G368MFD	Claribels	G920WAY	Worthern Travel
G213EOG	W M Buses	G274EOG	W M Buses	G373REG	Owen's	G929GWN	Chase
G214EOG	W M Buses	G275EOG	W M Buses	G373YUR	Zak's	GAX2C	Horrocks
G215EOG	W M Buses	G276EOG	W M Buses	G381JBD	Williamson	GBF73N	Cave
G215HCP	W M Buses	G277EOG	W M Buses	G399FSF	Midland	GBU7V	Stevensons
G216EOG	W M Buses	G278BEL	Bassetts	G417WFP	Stevensons	GCA747	Green Bus
G216HCP	W M Buses	G278EOG	W M Buses	G444JAW	Elcock Reisen	GDZ795	Stevensons
G217EOG	W M Buses	G279EOG	W M Buses	G453XHK	Britannia	GEU369N	Birmingham Coach
G217HCP	W M Buses	G279HDW	Blue Bus	G472EFA	Bassetts	GEU371N	Birmingham Coach
G218EOG	W M Buses	G280EOG	W M Buses	G477ERF	P M T	GFM101X	P M T
G218HCP	W M Buses	G281EOG	W M Buses	G478ERF	P M T	GFM102X	P M T
G219EOG	W M Buses	G282EOG	W M Buses	G495FFA	P M T	GFM103X	P M T
G220EOG	W M Buses	G283EOG	W M Buses	G505SFT	Midland	GFM104X	P M T
G221EOG	W M Buses	G284EOG	W M Buses	G507SFT	Midland	GFM105X	P M T
G222EOG	W M Buses	G285EOG	W M Buses	G510SFT	Midland	GFM106X	P M T
G223EOG	W M Buses	G286EOG	W M Buses	G511SFT	Midland	GFM107X	Midland
G224EOG	W M Buses	G287EOG	W M Buses	G529MNT	Jones	GFM108X	P M T
G225EOG	W M Buses	G288EOG	W M Buses	G532CVT	P M T	GFM109X	P M T
G226EOA	Zak's	G289EOG	W M Buses	G543JOG	W M Buses	GHE739V	Britannia
G226EOG	W M Buses	G290EOG	W M Buses	G550ERF	P M T	GHU641N	Birmingham Coach
G227EOA	Merry Hill	G291EOG	W M Buses	G587LUX	Jones	GHU645N	Ludlows
G227EOG	W M Buses	G292EOG	W M Buses	G594EKV	Travel de Courcey	GIL1909	Britannia
G228EOA	Merry Hill	G293EOG	W M Buses	G616WGS	Stevensons	GIL2195	Britannia
G228EOG	W M Buses	G294EOG	W M Buses	G644BPH	Midland	GIL2785	Boultons
G229EOG	W M Buses	G295EOG	W M Buses	G645BPH	Midland	GIL2942	W M Buses
G230EOG	W M Buses	G296EOG	W M Buses	G646BPH	Midland	GIL3167	Happy Days
G231EOG	W M Buses	G297EOG	W M Buses	G647BPH	Midland	GIL3273	Britannia
G232EOG	W M Buses	G298EOG	W M Buses	G675BFA	Blue Bus	GIL3274	Britannia
G233EOG	W M Buses	G299EOG	W M Buses	G678XVT	Blue Bus	GIL3275	Britannia
G234EOG	W M Buses	G300EOG	W M Buses	G701HOP	W M Buses	GLJ680N	Falcon Travel
G235EOG	W M Buses	G301DPA	Midland	G702HOP	W M Buses	GMA409N	Choice Travel
G236EOG	W M Buses	G301EOG	W M Buses	G703HOP	W M Buses	GMB372T	Midland
G237EOG	W M Buses	G302EOG	W M Buses	G704HOP	W M Buses	GMB373T	Midland
G238EOG	W M Buses	G303EOG	W M Buses	G705HOP	W M Buses	GMB374T	Midland
G239EOG	W M Buses	G304EOG	W M Buses	G706HOP	W M Buses	GMB376T	Midland
G240EOG	W M Buses	G305EOG	W M Buses	G707HOP	W M Buses	GMB377T	P M T

The North & West Midlands Bus Handbook

GMB378T	Midland	GOG161W	W M Buses	GOG236W	W M Buses	H153DJU	Warringtons
GMB383T	Midland	GOG162W	W M Buses	GOG237W	W M Buses	H153SKU	W M Buses
GMB390T	Midland	GOG163W	W M Buses	GOG238W	W M Buses	H154OEP	Worthern Travel
GMB650T	Birmingham Coach	GOG164W	W M Buses	GOG239W	W M Buses	H154SKU	W M Buses
GMB661T	Birmingham Coach	GOG165W	W M Buses	GOG240W	W M Buses	H155SKU	W M Buses
GMS295S	Stevensons	GOG166W	W M Buses	GOG241W	W M Buses	H156SKU	W M Buses
GNT434V	Britannia	GOG167W	W M Buses	GOG242W	W M Buses	H157EFK	Shropshire Ed
GNY432R	Green Bus	GOG168W	Stevensons	GOG243W	W M Buses	H157SKU	W M Buses
GOG91W	W M Buses	GOG168W	W M Buses	GOG244W	W M Buses	H158SKU	W M Buses
GOG92W	W M Buses	GOG169W	W M Buses	GOG245W	W M Buses	H159SKU	W M Buses
GOG93W	W M Buses	GOG170W	W M Buses	GOG246W	W M Buses	H160JRE	P M T
GOG94W	W M Buses	GOG171W	W M Buses	GOG247W	W M Buses	H166MFA	Stevensons
GOG95W	W M Buses	GOG172W	W M Buses	GOG248W	W M Buses	H176JVT	Stevensons
GOG96W	W M Buses	GOG173W	W M Buses	GOG249W	W M Buses	H177EJF	Britannia
GOG97W	W M Buses	GOG174W	W M Buses	GOG250W	W M Buses	H177JVT	Stevensons
GOG98W	W M Buses	GOG175W	W M Buses	GOG251W	W M Buses	H180JRE	P M T
GOG99W	W M Buses	GOG176W	W M Buses	GOG252W	W M Buses	H181DHA	Midland
GOG100W	W M Buses	GOG177W	W M Buses	GOG253W	W M Buses	H182DHA	Midland
GOG101W	W M Buses	GOG178W	W M Buses	GOG254W	W M Buses	H183DHA	Midland
GOG102W	W M Buses	GOG179W	W M Buses	GOG255W	W M Buses	H184DHA	Midland
GOG103W	W M Buses	GOG180W	W M Buses	GOG256W	W M Buses	H185DHA	Midland
GOG104W	W M Buses	GOG181W	W M Buses	GOG257W	W M Buses	H186EHA	Midland
GOG105W	W M Buses	GOG182W	W M Buses	GOG258W	W M Buses	H187EHA	Midland
GOG106W	W M Buses	GOG183W	W M Buses	GOG259W	W M Buses	H188EHA	Midland
GOG107W	W M Buses	GOG184W	W M Buses	GOG260W	W M Buses	H189CNS	P M T
GOG108W	W M Buses	GOG185W	W M Buses	GOG261W	W M Buses	H189EHA	Midland
GOG109W	W M Buses	GOG186W	W M Buses	GOG262W	W M Buses	H191EHA	Midland
GOG110W	W M Buses	GOG187W	W M Buses	GOG263W	W M Buses	H192JNF	Stevensons
GOG111W	W M Buses	GOG188W	W M Buses	GOG264W	W M Buses	H196JVT	Midland
GOG112W	W M Buses	GOG189W	W M Buses	GOG265W	W M Buses	H197JVT	Stevensons
GOG113W	W M Buses	GOG190W	W M Buses	GOG266W	W M Buses	H198JVT	Stevensons
GOG114W	W M Buses	GOG191W	W M Buses	GOG267W	W M Buses	H199KEH	Stevensons
GOG115W	W M Buses	GOG192W	W M Buses	GOG268W	W M Buses	H201LOM	W M Buses
GOG116W	W M Buses	GOG193W	W M Buses	GOG269W	W M Buses	H201LRF	Stevensons
GOG117W	W M Buses	GOG194W	W M Buses	GOG270W	W M Buses	H202JHP	P M T
GOG118W	W M Buses	GOG195W	W M Buses	GOG271W	W M Buses	H202LOM	W M Buses
GOG119W	W M Buses	GOG196W	W M Buses	GOG272W	Stevensons	H202LRF	Stevensons
GOG120W	W M Buses	GOG197W	W M Buses	GOG273W	W M Buses	H203JHP	P M T
GOG121W	W M Buses	GOG198W	W M Buses	GOG274W	W M Buses	H203LOM	W M Buses
GOG122W	W M Buses	GOG199W	W M Buses	GOG275W	W M Buses	H203TCP	W M Buses
GOG123W	W M Buses	GOG200W	W M Buses	GOG684N	Leon's	H204LOM	W M Buses
GOG124W	W M Buses	GOG201W	W M Buses	GOK518N	W M Buses	H206LOM	W M Buses
GOG125W	W M Buses	GOG202W	W M Buses	GRF701V	P M T	H207LOM	W M Buses
GOG126W	W M Buses	GOG203W	W M Buses	GRF704V	P M T	H208LOM	W M Buses
GOG127W	W M Buses	GOG204W	W M Buses	GRF706V	P M T	H209LOM	W M Buses
GOG128W	W M Buses	GOG205W	W M Buses	GRF707V	P M T	H210LOM	W M Buses
GOG129W	W M Buses	GOG206W	W M Buses	GRF708V	P M T	H211LOM	W M Buses
GOG130W	W M Buses	GOG207W	W M Buses	GRF709V	P M T	H212LOM	W M Buses
GOG131W	W M Buses	GOG208W	W M Buses	GRF710V	P M T	H215LOM	W M Buses
GOG132W	W M Buses	GOG209W	W M Buses	GRF711V	P M T	H217LOM	W M Buses
GOG133W	W M Buses	GOG210W	W M Buses	GRF714V	P M T	H218LOM	W M Buses
GOG134W	W M Buses	GOG211W	W M Buses	GRF715V	P M T	H219LOM	W M Buses
GOG135W	W M Buses	GOG212W	W M Buses	GRF716V	P M T	H220LOM	W M Buses
GOG136W	W M Buses	GOG213W	W M Buses	GRM351L	Birmingham Coach	H221LOM	W M Buses
GOG137W	W M Buses	GOG214W	W M Buses	GRM353L	Birmingham Coach	H223LOM	W M Buses
GOG138W	W M Buses	GOG215W	W M Buses	GSU7T	Boydon	H224LOM	W M Buses
GOG139W	W M Buses	GOG216W	W M Buses	GUG118N	Birmingham Coach	H225LOM	W M Buses
GOG140W	W M Buses	GOG217W	W M Buses	GWC31T	Williamson	H226LOM	W M Buses
GOG141W	W M Buses	GOG218W	W M Buses	GWP633N	Horrocks	H227LOM	W M Buses
GOG142W	W M Buses	GOG219W	W M Buses	GWY691N	Hi Ride	H228LOM	W M Buses
GOG143W	W M Buses	GOG220W	W M Buses	H2FTG	Flight's	H229LOM	W M Buses
GOG144W	W M Buses	GOG221W	W M Buses	H3FTG	Flight's	H231LOM	W M Buses
GOG145W	W M Buses	GOG222W	W M Buses	H10WLE	Birmingham Coach	H232LOM	W M Buses
GOG146W	W M Buses	GOG223W	Stevensons	H112DDS	Midland	H233LOM	W M Buses
GOG147W	W M Buses	GOG224W	W M Buses	H130MRW	W M Buses	H234LOM	W M Buses
GOG148W	W M Buses	GOG225W	W M Buses	H131CDB	Midland	H235LOM	W M Buses
GOG149W	W M Buses	GOG226W	W M Buses	H131MRW	W M Buses	H236LOM	W M Buses
GOG150W	W M Buses	GOG227W	W M Buses	H132CDB	Midland	H237LOM	W M Buses
GOG151W	W M Buses	GOG228W	W M Buses	H133CDB	Midland	H237RUX	Owen's
GOG152W	W M Buses	GOG229W	W M Buses	H134CDB	Midland	H238LOM	Shropshire Ed
GOG153W	W M Buses	GOG230W	W M Buses	H135CDB	Midland	H239LOM	W M Buses
GOG155W	W M Buses	GOG231W	W M Buses	H136CDB	Midland	H241LOM	W M Buses
GOG156W	W M Buses	GOG232W	W M Buses	H149SKU	W M Buses	H242LOM	W M Buses
GOG157W	W M Buses	GOG233W	W M Buses	H150SKU	W M Buses	H243LOM	W M Buses
GOG158W	W M Buses	GOG234W	W M Buses	H151SKU	W M Buses	H244LOM	W M Buses
GOG159W	W M Buses	GOG235W	W M Buses	H152SKU	W M Buses	H245LOM	W M Buses

H246LOM	W M Buses	H804GRE	P M T	HXI3010	Stevensons	J918SEH	P M T
H247LOM	W M Buses	H805AHA	Midland	HXI3011	Stevensons	J920HGD	P M T
H304HVT	Blue Bus	H805GRE	P M T	HXI3012	Stevensons	J921TUK	Cave
H313WUA	Stevensons	H806AHA	Midland	IDZ8561	Stevensons	J995GCP	W M Buses
H314WUA	Stevensons	H806GRE	P M T	IIL6436	Knotty	J996GCP	W M Buses
H329DHA	Midland	H807GRE	P M T	J1NCB	NCB Motors	J997GCP	W M Buses
H330DHA	Midland	H808GRE	P M T	J3KCB	Shropshire Ed	J997UAC	W M Buses
H330JVT	Leon's	H809GRE	P M T	J4FTG	Flight's	J998GCP	W M Buses
H331DHA	Midland	H834GLD	P M T	J4MMT	Bassetts	JAW84V	Elcock Reisen
H332DHA	Midland	H835GLD	P M T	J6FTG	Flight's	JAZ9870	Boultons
H334DHA	Midland	H836GLD	P M T	J24GCX	W M Buses	JBR690T	Ludlows
H335DHA	Midland	H851GRE	P M T	J25GCX	W M Buses	JBW527D	Boultons
H336DHA	Midland	H851NOC	Stevensons	J31SFA	Stevensons	JDZ4898	Birmingham Coach
H337DHA	Midland	H852GRE	P M T	J32SFA	Stevensons	JEA884N	Worthern Travel
H338DHA	Midland	H853GRE	P M T	J34GCX	W M Buses	JFD296V	Jones
H339DHA	Midland	H854GRE	P M T	J34SRF	Stevensons	JFV317S	North Birmingham
H351FDU	Shropshire Ed	H855GRE	P M T	J36SRF	Stevensons	JFV318S	North Birmingham
H351HRF	P M T	H856GRE	P M T	J37GCX	W M Buses	JFV319S	North Birmingham
H352HRF	P M T	H857GRE	P M T	J56GCX	Birmingham Coach	JFV320S	North Birmingham
H353HRF	P M T	H858GRE	P M T	J58GCX	W M Buses	JHA207L	Hi Ride
H354HVT	P M T	H859GRE	P M T	J62SRE	Leon's	JHA246L	Travel de Courcey
H355HVT	P M T	H860GRE	P M T	J65SRE	Leon's	JHE137W	Stevensons
H356HVT	P M T	H861GRE	P M T	J129VAW	Britannia	JHJ141V	Falcon Travel
H357HVT	P M T	H880NFS	Stevensons	J143SRF	Stevensons	JIL5227	Blue Bus
H358JRE	P M T	H922FEW	Bassetts	J162REH	Stevensons	JMB330T	Claribels
H359JRE	P M T	HCA969N	Chase	J169REH	Stevensons	JMB337T	Claribels
H361JRE	P M T	HDZ8350	W M Buses	J203REH	Stevensons	JNK984N	Jones
H362JRE	P M T	HDZ8352	W M Buses	J204REH	Stevensons	JNO52Y	Sandwell Travel
H363JRE	P M T	HFM183N	Chase	J205REH	Stevensons	JOX458P	Claribels
H366LFA	P M T	HHA115L	Falcon Travel	J206REH	Stevensons	JOX480P	Midland
H367LFA	P M T	HHA126L	Cave	J207REH	Stevensons	JOX496P	Serveverse
H368LFA	P M T	HHJ378Y	Travel de Courcey	J208SRF	Stevensons	JOX506P	Birmingham Coach
H369LFA	P M T	HHU632N	Chase	J209SRF	Stevensons	JOX530P	Choice Travel
H370LFA	P M T	HHU633N	Birmingham Coach	J272SOC	Ludlows	JOX717P	Midland
H371LFA	P M T	HHU634N	Birmingham Coach	J297KFP	Williamson	JPA171K	Knotty
H372MEH	P M T	HIL2375	Procters	J327VAW	Williamson	JPF103K	Knotty
H373MVT	P M T	HIL2376	Procters	J328RVT	P M T	JRB914N	Longmynd
H407LVC	W M Buses	HIL2377	Procters	J328VAW	Williamson	JRE354V	Bassetts
H408LVC	W M Buses	HIL2378	Procters	J348GKH	W M Buses	JTC993X	Little Red Bus
H408YMA	Stevensons	HIL2379	Procters	J405AWF	Bowens	JTH782P	Choice Travel
H433DHA	Midland	HIL3652	Stevensons	J407AWF	Bowens	JTH783P	Choice Travel
H481JRE	P M T	HIL6245	Leon's	J413NCP	Birmingham Coach	JTU228T	Boydon
H482JRE	P M T	HIL6584	Elcock Reisen	J414NCP	Birmingham Coach	JUH231W	Green Bus
H483JRE	P M T	HIL7386	Procters	J416UOC	M&J Travel	JUM531V	Stevensons
H483ONT	Shropshire Ed	HIL7613	Procters	J430WFA	P M T	JUS476Y	Minsterley
H484ONT	Shropshire Ed	HIL7614	Procters	J431WFA	P M T	JWF490W	Stevensons
H485ONT	Shropshire Ed	HIL7615	Procters	J436HDS	Jones	JWO891L	Knotty
H501GHA	Midland	HIL7616	Procters	J484PVT	P M T	K1HDC	Happy Days
H516YCX	W M Buses	HIL7620	Procters	J485PVT	P M T	K1NCB	NCB Motors
H517YCX	W M Buses	HIL7621	Procters	J486PVT	P M T	K1TGE	Longmynd
H552SWE	Metropolitan	HIL7622	Procters	J556GTP	Stevensons	K2KHW	Warringtons
H557XNN	Patterson	HIL7623	Procters	J648XHL	Metropolitan	K4GWC	Warringtons
H619FUT	Bowens	HIL7624	Procters	J701NHA	Midland	K6BUS	Bakers
H621FUT	Bowens	HIL8863	Butters	J733KBC	Warringtons	K7BUS	Bakers
H623FUT	Bowens	HJA131N	Birmingham Coach	J842NHN	Bassetts	K11HDC	Happy Days
H671ATN	Zak's	HKP126	King Offa	J844RAC	W M Buses	K12FTG	Flight's
H682FCU	Birmingham Coach	HLG812K	M&J Travel	J845RAC	W M Buses	K16FTG	Flight's
H708LOL	Midland	HMA565T	Birmingham Coach	J853TRW	W M Buses	K17FTG	Flight's
H709LOL	Midland	HMA658N	Chase	J901SEH	P M T	K18FTG	Flight's
H710LOL	Midland	HNB24N	Cave	J902SEH	P M T	K19FTG	Flight's
H713LOL	Merry Hill	HNL157N	Glenstuart	J903SEH	P M T	K20FTG	Flight's
H714LOL	Merry Hill	HPF298N	Rest & Ride	J904SEH	P M T	K106UFP	Owen's
H718LOX	W M Buses	HPN945	Zak's	J905SEH	P M T	K123CAW	Elcock Reisen
H719LOX	W M Buses	HRE128V	Procters	J906SEH	P M T	K136ARE	Stevensons
H720LOL	Ludlows	HRE129V	Procters	J907SEH	P M T	K137ARE	Stevensons
H720LOX	W M Buses	HSC106T	Birmingham Coach	J908SEH	P M T	K138BRF	Stevensons
H723LOL	Merry Hill	HST11	Harry Shaw	J909SEH	P M T	K139BRF	Stevensons
H729LOL	Midland	HTL755N	Handybus	J910SEH	P M T	K140BFA	Stevensons
H731AUE	Handybus	HUH409N	Chase	J911SEH	P M T	K140RYS	Stevensons
H731LOL	Midland	HUJ998V	Elcock Reisen	J912SEH	P M T	K141BFA	Stevensons
H761RNT	Williamson	HUJ999V	Elcock Reisen	J913SEH	P M T	K141RYS	Stevensons
H801GRE	P M T	HVJ203	Boultons	J914SEH	P M T	K142BFA	Stevensons
H802GRE	P M T	HXI3006	Stevensons	J915HGD	Shropshire Ed	K150BRF	Stevensons
H803AHA	Midland	HXI3007	Stevensons	J915SEH	P M T	K152DNT	Williamson
H803GRE	P M T	HXI3008	Stevensons	J916SEH	P M T	K153DNT	Williamson
H804AHA	Midland	HXI3009	Stevensons	J917SEH	P M T	K211UHA	Midland

K212UHA	Midland	KDZ5803	W M Buses	L1NCB	NCB Motors	L553LVT		P M T	
K213UHA	Midland	KDZ5804	W M Buses	L1NER	Flight's	L554LVT		P M T	
K214UHA	Midland	KDZ5805	W M Buses	L1NKF	Flight's	L556LVT		P M T	
K215UHA	Midland	KFM111Y	P M T	L2LCT	Leon's	L557LVT		P M T	
K216UHA	Midland	KFM112Y	P M T	L4WOL	Happy Days	L558LVT		P M T	
K217UHA	Midland	KFM113Y	P M T	L5URE	Harry Shaw	L605BNX		Midland	
K218UHA	Midland	KFM114Y	P M T	L19UER	Harry Shaw	L618BNX		Midland	
K219UHA	Midland	KFM115Y	P M T	L22FTG	Flight's	L620BNX		Midland	
K296GDT	Bowens	KIB6993	Travel de Courcey	L33FTG	Flight's	L767LAW		Shropshire Ed	
K297GDT	Bowens	KJW276W	W M Buses	L42VRW	Harry Shaw	L776LUJ		Owen's	
K298GDT	Bowens	KJW277W	W M Buses	L43VRW	Harry Shaw	L862HFA		P M T	
K299GDT	Bowens	KJW278W	W M Buses	L44FTG	Flight's	L930HFA		P M T	
K321AUX	Elcock Reisen	KJW279W	W M Buses	L53JUJ	M&J Travel	L931HFA		P M T	
K337ABH	Elcock Reisen	KJW280W	W M Buses	L55FTG	Flight's	L932HFA		P M T	
K374BRE	P M T	KJW281W	W M Buses	L64YJF	Owen's	L933HFA		P M T	
K375BRE	P M T	KJW282W	W M Buses	L66FTG	Flight's	L934HFA		P M T	
K432XRF	P M T	KJW283W	W M Buses	L77FTG	Flight's	L935HFA		P M T	
K433XRF	P M T	KJW284W	W M Buses	L94HRF	Stevensons	L936HFA		P M T	
K434XRF	P M T	KJW285W	W M Buses	L95HRF	Stevensons	L937LRF		P M T	
K435XRF	P M T	KJW286W	W M Buses	L100SBS	Stevensons	L938LRF		P M T	
K436XRF	P M T	KJW287W	W M Buses	L102MEH	Stevensons	L939LRF		P M T	
K437XRF	P M T	KJW288W	W M Buses	L226JFA	Stevensons	L940LRF		P M T	
K438XRF	P M T	KJW289W	W M Buses	L229JFA	Stevensons	L941LRF		P M T	
K439XRF	P M T	KJW290W	W M Buses	L230JFA	Stevensons	L942LRF		P M T	
K440DVT	Bassetts	KJW291W	W M Buses	L231NRE	P M T	L962NFA		Bassetts	
K440XRF	P M T	KJW292W	W M Buses	L232JFA	Stevensons	LFR875X		Midland	
K441XRF	P M T	KJW293W	W M Buses	L253NFA	Stevensons	LHS480P		Knotty	
K442XRF	P M T	KJW294W	W M Buses	L254NFA	Stevensons	LIL3059		Cave	
K443XRF	P M T	KJW295W	W M Buses	L255NFA	Stevensons	LIL7810		Britannia	
K445XRF	P M T	KJW296W	Stevensons	L269GBU	P M T	LIL7812		Britannia	
K446XRF	P M T	KJW297W	W M Buses	L300BVA	Happy Days	LIL7814		Britannia	
K447XRF	P M T	KJW298W	W M Buses	L300SBS	Stevensons	LIW1324		Rest & Ride	
K448XRF	P M T	KJW299W	W M Buses	L301NFA	Stevensons	LJI8160		Leon's	
K449XRF	P M T	KJW300W	W M Buses	L302NFA	Stevensons	LJN622P		W M Buses	
K487CVT	P M T	KJW301W	Stevensons	L303NFA	Stevensons	LMB948P		W M Buses	
K488CVT	P M T	KJW302W	W M Buses	L304NFA	Stevensons	LOA326X		W M Buses	
K489CVT	P M T	KJW303W	W M Buses	L305NFA	Stevensons	LOA327X		W M Buses	
K490CVT	P M T	KJW305W	Stevensons	L321BNX	Serveverse	LOA328X		W M Buses	
K491CVT	P M T	KJW306W	Stevensons	L321HRE	P M T	LOA329X		W M Buses	
K492CVT	P M T	KJW308W	W M Buses	L321JUJ	Elcock Reisen	LOA330X		W M Buses	
K544XRF	P M T	KJW309W	W M Buses	L323NRF	P M T	LOA331X		W M Buses	
K657BOH	Birmingham Coach	KJW310W	Stevensons	L405LHE	Bowens	LOA332X		W M Buses	
K658BOH	Birmingham Coach	KJW311W	W M Buses	L406LHE	Bowens	LOA333X		W M Buses	
K659BOH	Birmingham Coach	KJW312W	W M Buses	L407LHE	Bowens	LOA334X		W M Buses	
K660BOH	Birmingham Coach	KJW313W	W M Buses	L408LHE	Bowens	LOA335X		W M Buses	
K680BRE	Leon's	KJW314W	W M Buses	L455LVT	P M T	LOA336X		W M Buses	
K713RNR	Bowens	KJW315W	W M Buses	L493HRE	P M T	LOA337X		W M Buses	
K714RNR	Bowens	KJW316W	W M Buses	L494HRE	P M T	LOA338X		W M Buses	
K870ANT	Williamson	KJW317W	W M Buses	L495HRE	P M T	LOA339X		W M Buses	
K871ANT	Williamson	KJW318W	Stevensons	L496HRE	P M T	LOA340X		W M Buses	
K879EAW	Owen's	KJW319W	W M Buses	L497HRE	P M T	LOA341X		W M Buses	
K882DUJ	Elcock Reisen	KJW320W	Stevensons	L498HRE	P M T	LOA342X		W M Buses	
K892BSX	Leon's	KJW321W	W M Buses	L502BNX	Midland	LOA343X		W M Buses	
K916FVC	W M Buses	KJW322W	Stevensons	L503BNX	Midland	LOA344X		W M Buses	
K917FVC	W M Buses	KJW323W	W M Buses	L504BNX	Midland	LOA345X		W M Buses	
K918FVC	W M Buses	KJW324W	W M Buses	L506BNX	Midland	LOA346X		W M Buses	
K919XRF	P M T	KJW325W	W M Buses	L507BNX	Midland	LOA347X		W M Buses	
K920XRF	P M T	KMA401T	Midland	L508BNX	Midland	LOA348X		W M Buses	
K921XRF	P M T	KMA402T	Midland	L509BNX	Midland	LOA349X		W M Buses	
K922XRF	P M T	KNT780	Boultons	L510BNX	Midland	LOA350X		W M Buses	
K923XRF	P M T	KNV505P	Rest & Ride	L511BNX	Midland	LOA351X		W M Buses	
K924XRF	P M T	KNW656N	Chase	L512BNX	Midland	LOA352X		W M Buses	
K925XRF	P M T	KOM789P	Birmingham Coach	L513BNX	Midland	LOA353X		W M Buses	
K926XRF	P M T	KOM793P	Birmingham Coach	L514BNX	Midland	LOA354X		W M Buses	
K927XRF	P M T	KOV2	Harry Shaw	L515BNX	Midland	LOA355X		W M Buses	
K928XRF	P M T	KRN110T	Claribels	L516BNX	Midland	LOA356X		W M Buses	
K929XRF	P M T	KRN115T	Claribels	L517BNX	Midland	LOA357X		W M Buses	
K947BRE	Stevensons	KRO645T	Claribels	L519BNX	Midland	LOA358X		W M Buses	
KAD349V	Britannia	KRP562V	Cave	L521BNX	Midland	LOA359X		W M Buses	
KCR108P	Ludlows	KSO76P	Birmingham Coach	L522BNX	Midland	LOA360X		W M Buses	
KCW74N	Bassetts	KTA356V	Minsterley	L523BNX	Midland	LOA361X		W M Buses	
KCW75N	Bassetts	KUB671V	Stevensons	L540EHD	W M Buses	LOA362X		W M Buses	
KDM760T	NCB Motors	KUN497P	Boultons	L541EHD	W M Buses	LOA363X		W M Buses	
KDW360P	Cave	KVE906P	Boydon	L542EHD	W M Buses	LOA364X		W M Buses	
KDZ5801	W M Buses	KVE909P	Knotty	L543EHD	W M Buses	LOA365X		W M Buses	
KDZ5802	W M Buses	L1HDC	Happy Days	L545MRA	Warringtons	LOA366X		W M Buses	

LOA367X	W M Buses	LTG797P	Chase	M442BLC	N C P	M969XVT	P M T
LOA368X	W M Buses	LTG798P	Chase	M443BLC	N C P	M970XVT	P M T
LOA369X	W M Buses	LTG850P	Chase	M445BLC	N C P	M971XVT	P M T
LOA370X	W M Buses	LTS93X	Midland	M446BLC	N C P	M972XVT	P M T
LOA371X	W M Buses	LUG82P	Green Bus	M451EDH	Midland	MAW112P	Claribels
LOA372X	W M Buses	LUY742	Stevensons	M452EDH	Midland	MDL881R	Falcon Travel
LOA373X	W M Buses	LVS423V	Bassetts	M453EDH	Midland	MFA717V	P M T
LOA374X	W M Buses	M1FTG	Flight's	M454EDH	Midland	MFA718V	P M T
LOA375X	W M Buses	M1LCT	Leon's	M455EDH	Midland	MFA720V	P M T
LOA376X	W M Buses	M2FTG	Flight's	M456EDH	Midland	MFA723V	P M T
LOA377X	W M Buses	M2LCT	Leon's	M457EDH	Midland	MFR18P	Stevensons
LOA378X	W M Buses	M3FTG	Flight's	M458EDH	Midland	MFR41P	Stevensons
LOA379X	W M Buses	M7TUB	Choice Travel	M459EDH	Midland	MFR125P	Stevensons
LOA380X	W M Buses	M10FTG	Flight's	M460EDH	Midland	MFR126P	Stevensons
LOA381X	W M Buses	M20FTG	Flight's	M461EDH	Midland	MFV70P	Minsterley
LOA382X	W M Buses	M25YRE	P M T	M462EDH	Midland	MHJ722V	Stevensons
LOA383X	W M Buses	M26XEH	Choice Travel	M551ONT	King Offa	MHJ725V	Stevensons
LOA384X	W M Buses	M26YRE	P M T	M559SRE	P M T	MHJ727V	Stevensons
LOA385X	W M Buses	M27YRE	P M T	M561SRE	P M T	MHP17X	Minsterley
LOA386X	W M Buses	M28YRE	P M T	M562SRE	P M T	MIB104	Copeland's
LOA387X	W M Buses	M30FTG	Flight's	M563SRE	P M T	MIB116	Copeland's
LOA388X	Stevensons	M32LHP	Harry Shaw	M564SRE	P M T	MIB236	Copeland's
LOA389X	W M Buses	M34LHP	Harry Shaw	M565SRE	P M T	MIB246	Copeland's
LOA390X	W M Buses	M36LHP	Harry Shaw	M566SRE	P M T	MIB268	Copeland's
LOA391X	W M Buses	M37LHP	Harry Shaw	M567SRE	P M T	MIB279	Copeland's
LOA392X	W M Buses	M38LHP	Harry Shaw	M568SRE	P M T	MIB346	Copeland's
LOA393X	W M Buses	M40FTG	Flight's	M569SRE	P M T	MIB516	Copeland's
LOA394X	W M Buses	M46POL	Serveverse	M570SRE	P M T	MIB520	Copeland's
LOA395X	W M Buses	M50FTG	Flight's	M571BVL	Harry Shaw	MIB614	Copeland's
LOA396X	W M Buses	M60FTG	Flight's	M571SRE	P M T	MIB615	Copeland's
LOA397X	W M Buses	M75OUX	Owen's	M572SRE	P M T	MIB761	Copeland's
LOA398X	W M Buses	M101VKY	Boultons	M573SRE	P M T	MIB864	Copeland's
LOA399X	W M Buses	M123OUX	Elcock Reisen	M660SRE	P M T	MIB970	Copeland's
LOA400X	W M Buses	M123RAW	Elcock Reisen	M701HBC	Williamson	MIB4964	Boydon
LOA401X	W M Buses	M211FMR	Harry Shaw	M784SOF	Birmingham Coach	MIB9298	Travel de Courcey
LOA402X	W M Buses	M213NHP	Harry Shaw	M785SOF	Birmingham Coach	MJI2364	Travel de Courcey
LOA403X	W M Buses	M237SOJ	Birmingham Coach	M801OJW	Metropilitan	MJI2365	Travel de Courcey
LOA404X	W M Buses	M288OUR	Travel de Courcey	M802MOJ	Midland	MJI2366	Travel de Courcey
LOA405X	W M Buses	M291OUR	Travel de Courcey	M803MOJ	Midland	MJI2367	Travel de Courcey
LOA406X	W M Buses	M316VET	Bowens	M804MOJ	Midland	MJI2368	Travel de Courcey
LOA407X	W M Buses	M317LJW	W M Buses	M805MOJ	Midland	MJI2369	Travel de Courcey
LOA408X	W M Buses	M317VET	Bowens	M831SDA	Midland	MJI2370	Travel de Courcey
LOA409X	W M Buses	M318LJW	W M Buses	M832SDA	Midland	MJI4838	Travel de Courcey
LOA410X	W M Buses	M318VET	Bowens	M833SDA	Midland	MJI7861	Travel de Courcey
LOA411X	W M Buses	M319LJW	W M Buses	M834SDA	Midland	MJI7862	Travel de Courcey
LOA412X	W M Buses	M319VET	Bowens	M835SDA	Midland	MJI7863	Travel de Courcey
LOA413X	W M Buses	M320LJW	W M Buses	M845OKV	W M Buses	MLG961P	Choice Travel
LOA414X	W M Buses	M321LJW	W M Buses	M881WFA	Bassetts	MMB970P	Chase
LOA415X	W M Buses	M321RAW	Elcock Reisen	M882YEH	Bassetts	MMB973P	Glenstuart
LOA416X	W M Buses	M322LJW	W M Buses	M943SRE	P M T	MOD573P	Happy Days
LOA417X	W M Buses	M323LJW	W M Buses	M944SRE	P M T	MOD826P	Birmingham Coach
LOA418X	W M Buses	M324LJW	W M Buses	M944XET	Claribels	MOD850P	Birmingham Coach
LOA419X	W M Buses	M325LJW	W M Buses	M945SRE	P M T	MOF225	W M Buses
LOA420X	W M Buses	M326LJW	W M Buses	M946SRE	P M T	MRH162P	Clowes
LOA421X	W M Buses	M371EFD	Midland	M947SRE	P M T	MRP5V	Chase
LOA422X	W M Buses	M372EFD	Midland	M948SRE	P M T	MRY53P	NCB Motors
LOA423X	W M Buses	M373EFD	Midland	M949SRE	P M T	MTJ775S	Birmingham Coach
LOA424X	W M Buses	M374EFD	Midland	M951SRE	P M T	MTU120Y	P M T
LOA425X	W M Buses	M375EFD	Midland	M952SRE	P M T	MTU122Y	P M T
LOA426X	W M Buses	M376EFD	Midland	M953XVT	P M T	MTU123Y	P M T
LOA427X	W M Buses	M377EFD	Midland	M954HRY	Owen's	MTU124Y	P M T
LOA428X	W M Buses	M377SRE	P M T	M954XVT	P M T	MTU125Y	P M T
LOA429X	Stevensons	M378EFD	Midland	M955XVT	P M T	MUR215L	Worthern Travel
LOA430X	W M Buses	M378SRE	P M T	M956XVT	P M T	MUV837X	King Offa
LOA431X	W M Buses	M379EFD	Midland	M957XVT	P M T	MVS514	Worthern Travel
LOA432X	W M Buses	M379SRE	P M T	M958XVT	P M T	MWA839P	Knotty
LOA433X	W M Buses	M380EFD	Midland	M959XVT	P M T	MWB115P	Minsterley
LOA434X	W M Buses	M380SRE	P M T	M960XVT	P M T	MWJ730W	Patterson
LOA435X	W M Buses	M381EFD	Midland	M961XVT	P M T	N2LCT	Leon's
LOI1454	Leon's	M381SRE	P M T	M962XVT	P M T	N3LCT	Leon's
LOI7191	Leon's	M382SRE	P M T	M963XVT	P M T	N11HDC	Happy Days
LOI8643	Leon's	M383SRE	P M T	M964XVT	P M T	N23FWU	W M Buses
LOI9772	Leon's	M401EFD	Midland	M965XVT	P M T	N31EVT	Choice Travel
LPR940P	Birmingham Coach	M402EFD	Midland	M966XVT	P M T	N32EVT	Choice Travel
LRC21W	Longmynd	M403EFD	Midland	M967XVT	P M T	N53FWU	W M Buses
LTG796P	Chase	M404EFD	Midland	M968XVT	P M T	N54FWU	W M Buses

The North & West Midlands Bus Handbook   **127**

N63FWU	Happy Days	N371WOH	W M Buses	N811XOJ	Merry Hill	NOA470X	W M Buses
N91WOM	Merry Hill	N372WOH	W M Buses	N812XOJ	Merry Hill	NOA471X	W M Buses
N91WVC	Harry Shaw	N373WOH	W M Buses	N863CEH	P M T	NOA472X	W M Buses
N92WOM	Merry Hill	N374WOH	W M Buses	N864CEH	P M T	NOA473X	W M Buses
N92WVC	Harry Shaw	N375WOH	W M Buses	N865CEH	P M T	NOA474X	W M Buses
N93WOM	Merry Hill	N376WOH	W M Buses	N866CEH	P M T	NOA475X	W M Buses
N93WVC	Harry Shaw	N377WOH	W M Buses	N867CEH	P M T	NOC434R	W M Buses
N94WOM	Merry Hill	N378WOH	W M Buses	N986TWK	W M Buses	NOC444R	W M Buses
N94WVC	Harry Shaw	N405HVT	P M T	NAT222A	Ludlows	NOC445R	W M Buses
N95WOM	Merry Hill	N406HVT	P M T	NAT333A	Ludlows	NOC471R	W M Buses
N95WVC	Harry Shaw	N407HVT	P M T	NAT555A	Ludlows	NOC477R	W M Buses
N96WVC	Harry Shaw	N408HVT	P M T	NBF957V	Handybus	NOC488R	W M Buses
N97WVC	Harry Shaw	N409HVT	P M T	NBZ1676	Green Bus	NOC600R	W M Buses
N98WVC	Harry Shaw	N410HVT	P M T	NCW151T	Green Bus	NOC610R	W M Buses
N110VAW	M&J Travel	N411HVT	P M T	NCW152T	Green Bus	NOC615R	W M Buses
N133GRF	Choice Travel	N412HVT	P M T	NDH7P	Leon's	NOC723R	W M Buses
N133WOH	W M Buses	N413HVT	P M T	NDT634X	Zak's	NOC728R	W M Buses
N147BOF	Merry Hill	N414HVT	P M T	NED433W	P M T	NOC732R	W M Buses
N148BOF	Merry Hill	N415HVT	P M T	NEH725W	P M T	NOC735R	W M Buses
N149BOF	Merry Hill	N416HVT	P M T	NEH727W	P M T	NOC741R	W M Buses
N151BOF	Merry Hill	N417HVT	P M T	NEH728W	P M T	NOC744R	W M Buses
N152BOF	Merry Hill	N418HVT	P M T	NEH729W	P M T	NOE567R	Birmingham Coach
N153BOF	Merry Hill	N419HVT	P M T	NEH731W	P M T	NOE568R	Birmingham Coach
N154BOF	Merry Hill	N463EHA	Midland	NEH732W	P M T	NOE570R	Birmingham Coach
N155BOF	Merry Hill	N464EHA	Midland	NEL111P	Claribels	NOE575R	Midland
N156BOF	Merry Hill	N465EHA	Midland	NEL860M	Birmingham Coach	NOE579R	Birmingham Coach
N157BOF	Merry Hill	N466EHA	Midland	NEN957R	Chase	NPD111L	Ludlows
N268XOJ	Merry Hill	N467EHA	Midland	NEN958R	Falcon Travel	NTC604R	Birmingham Coach
N269XOJ	Merry Hill	N468EHA	Midland	NEN965R	Chase	NTC606R	W M Buses
N270XOJ	Merry Hill	N468WDA	Merry Hill	NFN69M	Falcon Travel	NTC609R	W M Buses
N301ENX	Midland	N469EHA	Midland	NFW966P	Rest & Ride	NTC612R	W M Buses
N302ENX	Midland	N469WDA	Merry Hill	NIW2317	Britannia	NTC614R	Birmingham Coach
N303ENX	Midland	N470EHA	Midland	NIW2318	Britannia	NTC619R	W M Buses
N304ENX	Midland	N471EHA	Midland	NIW2320	Britannia	NTC623R	W M Buses
N305ENX	Midland	N472EHA	Midland	NIW2321	Britannia	NTC625R	Stevensons
N329WOH	W M Buses	N495TVP	Flight's	NIW2322	Britannia	NTX576R	Green Bus
N331WOH	W M Buses	N574CEH	P M T	NIW3546	Britannia	NUS6P	Boydon
N331WOH	W M Buses	N575CEH	P M T	NIW5986	Boydon	NVR907W	Longmynd
N332WOH	W M Buses	N576CEH	P M T	NIW6131	Britannia	NWK10P	Minsterley
N334WOH	W M Buses	N577CEH	P M T	NKN101M	Knotty .	NWO460R	Ludlows
N335WOH	W M Buses	N578CEH	P M T	NLC871V	Jones	NWO467R	Chase
N336WOH	W M Buses	N579CEH	P M T	NOA199R	Birmingham Coach	NWO476R	Cave
N337WOH	W M Buses	N580CEH	P M T	NOA436X	W M Buses	NWO483R	Cave
N338WOH	W M Buses	N581CEH	P M T	NOA437X	W M Buses	NWO493R	Cave
N339WOH	W M Buses	N582CEH	P M T	NOA438X	W M Buses	NWO499R	Cave
N340WOH	W M Buses	N583CEH	P M T	NOA439X	W M Buses	NWW163K	Bassetts
N341WOH	W M Buses	N584CEH	P M T	NOA440X	W M Buses	OAF40M	Sandwell Travel
N342WOH	W M Buses	N585CEH	P M T	NOA441X	W M Buses	OAO564M	Falcon Travel
N343WOH	W M Buses	N586CEH	P M T	NOA442X	W M Buses	OCN423X	Rest & Ride
N344WOH	W M Buses	N587CEH	P M T	NOA443X	W M Buses	ODD161M	Boultons
N345WOH	W M Buses	N588CEH	P M T	NOA444X	W M Buses	ODM409V	P M T
N346WOH	W M Buses	N589CEH	P M T	NOA445X	W M Buses	ODM501V	Claribels
N347WOH	W M Buses	N590CEH	P M T	NOA446X	W M Buses	OFD231P	Ludlows
N348WOH	W M Buses	N591CEH	P M T	NOA447X	W M Buses	OFR983M	Knotty
N349WOH	W M Buses	N592CEH	P M T	NOA448X	W M Buses	OGL518	Stevensons
N350WOH	W M Buses	N593CEH	P M T	NOA449X	W M Buses	OIA1652	Leon's
N351WOH	W M Buses	N594CEH	P M T	NOA450X	W M Buses	OIJ6726	Worthern Travel
N352WOH	W M Buses	N627BWG	Lionspeed/Pete's	NOA451X	W M Buses	OIW5807	Britannia
N353WOH	W M Buses	N628BWG	Lionspeed/Pete's	NOA452X	W M Buses	OIW7023	Britannia
N354WOH	W M Buses	N629BWG	Lionspeed/Pete's	NOA453X	W M Buses	OIW7026	Britannia
N355WOH	W M Buses	N683AHL	Birmingham Coach	NOA454X	W M Buses	OIW7027	Britannia
N356WOH	W M Buses	N684AHL	Birmingham Coach	NOA455X	W M Buses	OIW7115	Britannia
N357WOH	W M Buses	N755GBF	Bassetts	NOA456X	W M Buses	OJD60R	Hi Ride
N358WOH	W M Buses	N782EHA	Midland	NOA457X	W M Buses	OJD129R	Travel de Courcey
N359WOH	W M Buses	N783EHA	Midland	NOA458X	W M Buses	OJD207R	Travel de Courcey
N360WOH	W M Buses	N784EHA	Midland	NOA459X	W M Buses	OJD210R	Travel de Courcey
N361WOH	W M Buses	N785EHA	Midland	NOA460X	W M Buses	OJD215R	Travel de Courcey
N362WOH	W M Buses	N786EHA	Midland	NOA461X	W M Buses	OJD220R	Travel de Courcey
N363WOH	W M Buses	N787EHA	Midland	NOA462X	W M Buses	OJD858R	Chase
N364WOH	W M Buses	N788EHA	Midland	NOA463X	W M Buses	OJD862R	Birmingham Coach
N365WOH	W M Buses	N789EHA	Midland	NOA464X	W M Buses	OJD863R	Chase
N366WOH	W M Buses	N790EHA	Midland	NOA465X	W M Buses	OJD865R	Chase
N367WOH	W M Buses	N791EHA	Midland	NOA466X	W M Buses	OJD868R	Chase
N368WOH	W M Buses	N806EHA	Midland	NOA467X	W M Buses	OJD870R	Chase
N369WOH	W M Buses	N807EHA	Midland	NOA468X	W M Buses	OKY822X	Stevensons
N370WOH	W M Buses	N808EHA	Midland	NOA469X	W M Buses	ONL924M	Minsterley

A rare survivor of the London Country SMA class of AEC Swifts is Knotty 23, JPF103K introduced for the 725 route in 1972 and seen here destined for Burslem. This Newcastle-based operator has the only licenced example of the type with one other vehicle known to be preserved. *Cliff Beeton*

OOU534M	Butters	POG478Y	W M Buses	POG513Y	W M Buses	POG548Y	W M Buses
OOX802R	W M Buses	POG479Y	W M Buses	POG514Y	W M Buses	POG549Y	W M Buses
OOX804R	W M Buses	POG480Y	W M Buses	POG515Y	W M Buses	POG550Y	W M Buses
OOX805R	W M Buses	POG481Y	W M Buses	POG516Y	W M Buses	POG551Y	W M Buses
OOX806R	W M Buses	POG482Y	W M Buses	POG517Y	W M Buses	POG552Y	W M Buses
OOX807R	W M Buses	POG483Y	W M Buses	POG518Y	W M Buses	POG553Y	W M Buses
OOX808R	W M Buses	POG484Y	W M Buses	POG519Y	W M Buses	POG554Y	W M Buses
OOX810R	W M Buses	POG485Y	W M Buses	POG520Y	W M Buses	POG555Y	W M Buses
OOX812R	W M Buses	POG486Y	W M Buses	POG521Y	W M Buses	POG556Y	W M Buses
OOX813R	W M Buses	POG487Y	W M Buses	POG522Y	W M Buses	POG557Y	W M Buses
OOX818R	W M Buses	POG488Y	W M Buses	POG523Y	W M Buses	POG558Y	W M Buses
OOX820R	Serveverse	POG489Y	W M Buses	POG524Y	W M Buses	POG559Y	W M Buses
OOX821R	W M Buses	POG490Y	W M Buses	POG525Y	W M Buses	POG560Y	W M Buses
OOX822R	W M Buses	POG491Y	W M Buses	POG526Y	W M Buses	POG561Y	W M Buses
OOX825R	W M Buses	POG492Y	W M Buses	POG527Y	W M Buses	POG562Y	W M Buses
ORP466M	Ludlows	POG493Y	W M Buses	POG528Y	W M Buses	POG563Y	W M Buses
OSR204R	Happy Days	POG494Y	W M Buses	POG529Y	W M Buses	POG564Y	W M Buses
OVT344P	Stevensons	POG495Y	W M Buses	POG530Y	W M Buses	POG565Y	W M Buses
PCA420V	P M T	POG496Y	W M Buses	POG531Y	W M Buses	POG566Y	W M Buses
PCA421V	P M T	POG497Y	W M Buses	POG532Y	W M Buses	POG567Y	W M Buses
PCD74R	Choice Travel	POG498Y	W M Buses	POG533Y	W M Buses	POG568Y	W M Buses
PCD75R	Rest & Ride	POG499Y	W M Buses	POG534Y	W M Buses	POG569Y	W M Buses
PCN423M	Rest & Ride	POG500Y	W M Buses	POG535Y	W M Buses	POG570Y	W M Buses
PCW946	Stevensons	POG501Y	W M Buses	POG536Y	W M Buses	POG571Y	W M Buses
PDK308S	NCB Motors	POG502Y	W M Buses	POG537Y	W M Buses	POG572Y	W M Buses
PDN405P	Zak's	POG503Y	W M Buses	POG538Y	W M Buses	POG573Y	W M Buses
PFM130Y	Midland	POG504Y	W M Buses	POG539Y	W M Buses	POG574Y	W M Buses
PGX235L	Horrocks	POG505Y	W M Buses	POG540Y	W M Buses	POG575Y	W M Buses
PHH613R	Claribels	POG506Y	W M Buses	POG541Y	W M Buses	POG576Y	W M Buses
PJP933R	Worthern Travel	POG507Y	W M Buses	POG542Y	W M Buses	POG577Y	W M Buses
PJT205R	M&J Travel	POG508Y	W M Buses	POG543Y	W M Buses	POG578Y	W M Buses
PJT256R	Choice Travel	POG509Y	W M Buses	POG544Y	W M Buses	POG579Y	W M Buses
PNH184	Leon's	POG510Y	W M Buses	POG545Y	W M Buses	POG580Y	W M Buses
POG476Y	W M Buses	POG511Y	W M Buses	POG546Y	W M Buses	POG581Y	W M Buses
POG477Y	W M Buses	POG512Y	W M Buses	POG547Y	W M Buses	POG582Y	W M Buses

The North & West Midlands Bus Handbook

POG579Y	W M Buses	RLG430V	P M T	SDA628S	W M Buses	THX222S	Chase
POG580Y	W M Buses	RMA443V	P M T	SDA630S	W M Buses	THX236S	Chase
POG581Y	W M Buses	ROK470M	W M Buses	SDA634S	W M Buses	THX260S	Chase
POG582Y	W M Buses	ROX611Y	W M Buses	SDA639S	W M Buses	THX264S	Chase
POG583Y	W M Buses	ROX612Y	W M Buses	SDA642S	W M Buses	THX266S	Chase
POG584Y	W M Buses	ROX613Y	W M Buses	SDA643S	W M Buses	TIA5734	W M Buses
POG585Y	W M Buses	ROX614Y	W M Buses	SDA646S	W M Buses	TIB2865	Boydon
POG586Y	W M Buses	ROX615Y	W M Buses	SDA648S	W M Buses	TIB2893	Knotty
POG587Y	W M Buses	ROX616Y	W M Buses	SDA649S	W M Buses	TJN509R	Choice Travel
POG588Y	W M Buses	ROX617Y	W M Buses	SDA650S	W M Buses	TMB877R	Green Bus
POG589Y	W M Buses	ROX618Y	W M Buses	SDA660S	W M Buses	TMB878R	Green Bus
POG590Y	W M Buses	ROX619Y	W M Buses	SDA699S	W M Buses	TMB879R	Green Bus
POG591Y	W M Buses	ROX620Y	W M Buses	SDA709S	W M Buses	TMJ633R	Butters
POG592Y	W M Buses	ROX621Y	W M Buses	SDA710S	W M Buses	TOB377	Flight's
POG593Y	W M Buses	ROX622Y	W M Buses	SDA712S	W M Buses	TOE477N	W M Buses
POG594Y	W M Buses	ROX623Y	W M Buses	SDA714S	W M Buses	TOE478N	W M Buses
POG595Y	W M Buses	ROX624Y	W M Buses	SDA718S	W M Buses	TOE485N	W M Buses
POG596Y	W M Buses	ROX625Y	W M Buses	SDA752S	W M Buses	TOE499N	W M Buses
POG597Y	W M Buses	ROX626Y	W M Buses	SDA757S	W M Buses	TOE502N	W M Buses
POG598Y	W M Buses	ROX627Y	W M Buses	SDA760S	W M Buses	TOE507N	W M Buses
POG599Y	W M Buses	ROX628Y	W M Buses	SDA764S	W M Buses	TOE509N	W M Buses
POG600Y	W M Buses	ROX629Y	W M Buses	SDA767S	W M Buses	TOE517N	W M Buses
POG601Y	W M Buses	ROX630Y	W M Buses	SDA772S	W M Buses	TOF683S	Britannia
POG602Y	W M Buses	ROX631Y	W M Buses	SDA800S	W M Buses	TOF684S	Midland
POG603Y	W M Buses	ROX632Y	W M Buses	SDW236Y	Sandwell Travel	TOF685S	Midland
POG604Y	W M Buses	ROX633Y	W M Buses	SEL236N	Rest & Ride	TOF687S	Midland
POG605Y	W M Buses	ROX634Y	W M Buses	SEL530X	Sandwell Travel	TOF690S	Midland
POG606Y	W M Buses	ROX635Y	W M Buses	SFA287R	P M T	TOF691S	Choice Travel
POG607Y	W M Buses	ROX636Y	W M Buses	SFJ139R	Choice Travel	TOF692S	Midland
POG608Y	W M Buses	ROX637Y	W M Buses	SFP829X	Warringtons	TOF693S	Midland
POG609Y	W M Buses	ROX638Y	W M Buses	SGR559R	W M Buses	TOF697S	Midland
POG610Y	W M Buses	ROX639Y	W M Buses	SGR565R	Falcon Travel	TOF698S	Midland
PPH431R	Knotty	ROX640Y	W M Buses	SIB3053	Boydon	TOF699S	Midland
PPM894R	Chase	ROX641Y	W M Buses	SIB3415	Boydon	TOF700S	Midland
PSU606	Travel de Courcey	ROX642Y	W M Buses	SIB6719	Happy Days	TOF701S	Midland
PSU906	Chase	ROX643Y	W M Buses	SIB7882	Boydon	TOF702S	Midland
PSU942	Chase	ROX644Y	W M Buses	SIB8342	Birmingham Coach	TOF703S	Midland
PSU946	Chase	ROX645Y	W M Buses	SJI5615	Rest & Ride	TOF704S	Midland
PSU954	Chase	ROX646Y	W M Buses	SKF13T	Birmingham Coach	TOF705S	Midland
PSU969	Chase	ROX647Y	W M Buses	SKF17T	Birmingham Coach	TOF718S	Midland
PSU977	Chase	ROX648Y	W M Buses	SKF18T	Birmingham Coach	TOF719S	Midland
PSU987	Chase	ROX649Y	W M Buses	SKF20T	Birmingham Coach	TOJ592S	Stevensons
PSU988	Chase	ROX650Y	W M Buses	SKF30T	Birmingham Coach	TOU962	Stevensons
PSU989	Chase	ROX651Y	W M Buses	SMU924N	Travel de Courcey	TPC101X	Midland
PSV323	Stevensons	ROX652Y	W M Buses	SNM71R	Minsterley	TPC102X	Midland
PTF730L	Birmingham Coach	ROX653Y	W M Buses	SOA673S	Claribels	TPC103X	Midland
PTF745L	W M Buses	ROX654Y	W M Buses	SPC287R	Chase	TPC104X	Midland
PTF751L	Choice Travel	ROX655Y	W M Buses	STJ31T	W M Buses	TPC107X	Midland
PTF758L	W M Buses	ROX656Y	W M Buses	STJ34T	W M Buses	TPC114X	Midland
PTT80R	Ludlows	ROX657Y	W M Buses	STK124T	North Birmingham	TPD12S	Knotty
PUA315W	Stevensons	ROX658Y	W M Buses	STK125T	North Birmingham	TPD194M	Ludlows
PUJ925	Elcock Reisen	ROX659Y	W M Buses	STK129T	North Birmingham	TPD195M	Ludlows
PUK633R	W M Buses	ROX660Y	W M Buses	STK130T	North Birmingham	TPE159S	Midland
PUK634R	W M Buses	ROX661Y	W M Buses	STK131T	North Birmingham	TPE163S	Midland
PUK637R	Midland	ROX663Y	W M Buses	STW18R	Stevensons	TPE166S	Midland
PUK639R	Midland	ROX664Y	W M Buses	STW20W	Stevensons	TPJ282S	Williamson
PUK647R	Midland	ROX665Y	W M Buses	SUA121R	Green Bus	TR6147	Midland
PUK652R	Midland	ROX666Y	W M Buses	SUX260R	Happy Days	TRN807U	Chase
PUK653R	Midland	ROX667Y	W M Buses	SXD696	Ludlows	TRN808V	Chase
PVT244L	Ludlows	RRR520R	Minsterley	TCH275L	Sandwell Travel	TRN809X	Chase
PYG139R	Minsterley	RTE111G	Chase	TDC829X	Sandwell Travel	TTC534T	Cave
Q246FVT	Stevensons	RWM582T	Copeland's	TDC854X	Midland	TTC539T	Birmingham Coach
RAW32R	Clowes	RYG385R	Happy Days	TDF224R	Sandwell Travel	TTT236R	Horrocks
RAW735X	Jones	RYG389R	Happy Days	TEH377W	Leon's	TUB7M	Choice Travel
RAX806M	Minsterley	RYG768R	Birmingham Coach	TEL493R	Chase	TUP432R	Cave
RFM886R	Ludlows	SBD525R	Choice Travel	TFG221X	Butters	TVP836S	W M Buses
RFM887M	Glenstaff	SCK708P	Birmingham Coach	TGE93	Longmynd	TVP837S	W M Buses
RFR177R	Bassetts	SDA515S	W M Buses	THX117S	Chase	TVP838S	W M Buses
RGS93R	Williamson	SDA537S	W M Buses	THX149S	Chase	TVP839S	W M Buses
RIB3524	Clowes	SDA538S	W M Buses	THX151S	Chase	TVP840S	W M Buses
RIB8034	Boydon	SDA545S	W M Buses	THX159S	Chase	TVP841S	W M Buses
RKA870T	Chase	SDA550S	W M Buses	THX160S	Chase	TVP842S	W M Buses
RKA873T	Birmingham Coach	SDA557S	W M Buses	THX181S	Chase	TVP843S	W M Buses
RKA877T	Birmingham Coach	SDA619S	W M Buses	THX193S	Chase	TVP846S	W M Buses
RKA878T	Birmingham Coach	SDA621S	W M Buses	THX209S	Chase	TVP847S	W M Buses
RKA884T	Birmingham Coach	SDA624S	W M Buses	THX219S	Birmingham Coach	TVP848S	W M Buses

The North & West Midlands Bus Handbook

TVP850S	W M Buses	VKE567S	Rest & Ride	WDA951T	W M Buses	WTU488W	P M T
TVP854S	W M Buses	VNO731S	W M Buses	WDA952T	W M Buses	WTU489W	P M T
TVP856S	W M Buses	VOD545K	Horrocks	WDA954T	W M Buses	WTU491W	P M T
TVP862S	W M Buses	VOI6874	Stevensons	WDA955T	W M Buses	WVH868V	Bassetts
TVP863S	W M Buses	VPT596R	Ludlows	WDA956T	W M Buses	WVT900S	P M T
TVP864S	W M Buses	VPT945R	Rest & Ride	WDA957T	W M Buses	WWA299Y	Claribels
TVP865S	W M Buses	VRF566X	Leon's	WDA958T	W M Buses	WWR424L	Boydon
TVP866S	W M Buses	VRF660S	Handybus	WDA960T	W M Buses	WYJ166S	Birmingham Coach
TVP871S	W M Buses	VRY1X	Clowes	WDA961T	W M Buses	WYJ167S	Birmingham Coach
TVP872S	W M Buses	VVH861V	Boydon	WDA962T	W M Buses	WYR562	Stevensons
TVP873S	W M Buses	VWX352X	Butters	WDA964T	W M Buses	XAF759	Stevensons
TVP874S	W M Buses	WAW356S	Elcock Reisen	WDA965T	W M Buses	XAK451T	Birmingham Coach
TVP875S	W M Buses	WBN464T	Chase	WDA966T	W M Buses	XBF423X	Bassetts
TVP876S	W M Buses	WBN478T	Falcon Travel	WDA967T	W M Buses	XBX831T	Boydon
TVP881S	W M Buses	WCJ500T	Travel de Courcey	WDA968T	W M Buses	XDF7X	Elcock Reisen
TVP885S	W M Buses	WCK128V	Procters	WDA969T	W M Buses	XGS764X	Bassetts
TVP888S	W M Buses	WCK132V	Claribels	WDA970T	W M Buses	XHA875	Little Red Bus
TVP889S	W M Buses	WDA661T	W M Buses	WDA971T	W M Buses	XIB3102	Rest & Ride
TVP890S	W M Buses	WDA662T	W M Buses	WDA972T	W M Buses	XIB3104	Rest & Ride
TVP891S	W M Buses	WDA663T	W M Buses	WDA973T	W M Buses	XIB3106	Rest & Ride
TVP895S	W M Buses	WDA665T	W M Buses	WDA976T	W M Buses	XLD627	Choice Travel
TVP897S	W M Buses	WDA666T	W M Buses	WDA977T	W M Buses	XNG773S	Falcon Travel
TVP898S	W M Buses	WDA667T	W M Buses	WDA978T	W M Buses	XOR841	Stevensons
TVP901S	W M Buses	WDA669T	W M Buses	WDA979T	W M Buses	XPD230N	Rest & Ride
TVP902S	W M Buses	WDA670T	W M Buses	WDA982T	W M Buses	XRF1X	P M T
TVP904S	W M Buses	WDA672T	W M Buses	WDA983T	W M Buses	XRF2X	P M T
UAB943Y	Longmynd	WDA673T	W M Buses	WDA984T	W M Buses	XRR584M	Falcon Travel
UAD316H	Knotty	WDA677T	W M Buses	WDA985T	W M Buses	XWG628T	Green Bus
UBW788	Williamson	WDA681T	W M Buses	WDA986T	W M Buses	XWJ791T	Jones
UDM450V	P M T	WDA682T	W M Buses	WDA987T	W M Buses	XWX175S	Happy Days
UET678S	Green Bus	WDA686T	W M Buses	WDA988T	W M Buses	XWX176S	Knotty
UFG55S	Birmingham Coach	WDA688T	W M Buses	WDA989T	W M Buses	YBF685S	P M T
UFG57S	Birmingham Coach	WDA689T	W M Buses	WDA991T	W M Buses	YBJ403	Chase
UFG59S	Birmingham Coach	WDA690T	W M Buses	WDA996T	W M Buses	YBO17T	Green Bus
UHG725R	W M Buses	WDA700T	W M Buses	WDA998T	W M Buses	YBW487V	P M T
UHG734R	W M Buses	WDA835T	W M Buses	WDA999T	W M Buses	YBW489V	P M T
UHG740R	W M Buses	WDA906T	W M Buses	WEB407T	Knotty	YCD79T	Birmingham Coach
UHG745R	Birmingham Coach	WDA907T	W M Buses	WEB408T	Knotty	YCD80T	Birmingham Coach
UHG748R	Rest & Ride	WDA909T	W M Buses	WEB410T	Knotty	YCD83T	Serveverse
UIB4589	Boydon	WDA911T	W M Buses	WEB411T	Knotty	YCD84T	Serveverse
UJN430Y	Midland	WDA912T	W M Buses	WFH169S	Travel de Courcey	YCW843N	Birmingham Coach
UMB333R	P M T	WDA913T	W M Buses	WJH503Y	Minsterley	YCW845N	Birmingham Coach
UNW929R	Happy Days	WDA915T	W M Buses	WLO574G	Horrocks	YEV310S	Rest & Ride
UOI772	Stevensons	WDA916T	W M Buses	WLT702	W M Buses	YHA116	Little Red Bus
UPE199M	Ludlows	WDA918T	W M Buses	WNO556L	Ludlows	YOK69K	W M Buses
UPE204M	Glenstuart	WDA919T	W M Buses	WNO558L	Falcon Travel	YPF773T	Chase
URF674S	P M T	WDA920T	W M Buses	WNO559L	Serveverse	YPF774T	Birmingham Coach
URN153V	Green Bus	WDA922T	W M Buses	WNR606S	Jones	YPJ207Y	Midland
URN321V	North Birmingham	WDA923T	W M Buses	WOC731T	Choice Travel	YPL397T	Cave
UTU981R	Rest & Ride	WDA924T	W M Buses	WOC739T	King Offa	YRE465S	Happy Days
UTV224S	Leon's	WDA925T	W M Buses	WPG224M	Birmingham Coach	YSF85S	Stevensons
UUX842S	Butters	WDA926T	W M Buses	WPH118Y	Midland	YSU953	Stevensons
UWJ628S	Butters	WDA928T	W M Buses	WPH121Y	Midland	YSU954	Stevensons
UWW512X	Stevensons	WDA930T	W M Buses	WPH122Y	Midland	YUT326T	Procters
UWW513X	Stevensons	WDA931T	W M Buses	WPH123Y	Midland	YWD687	Ludlows
UWW515X	Stevensons	WDA932T	W M Buses	WPH125Y	Midland	YXI6246	Clowes
UWW517X	Stevensons	WDA933T	W M Buses	WPH126Y	Midland	YXI6366	Clowes
VAJ785S	Stevensons	WDA934T	W M Buses	WPH139Y	Midland	YXI6367	Clowes
VCA452W	P M T	WDA935T	W M Buses	WTJ905L	Green Bus	YXI7340	Clowes
VCA458W	Midland	WDA940T	W M Buses	WTU465W	P M T	YYE270T	Chase
VCA460W	Midland	WDA941T	W M Buses	WTU470W	Midland	YYE274T	Chase
VCA464W	P M T	WDA942T	W M Buses	WTU472W	P M T	YYE291T	Birmingham Coach
VFV8V	Happy Days	WDA945T	W M Buses	WTU481W	P M T	YYE295T	Chase
VFX983S	Falcon Travel	WDA947T	W M Buses	WTU482W	P M T	YYJ955	Stevensons
VIB6165	Boydon	WDA949T	W M Buses	WTU483W	P M T		
VJT607X	Bassetts	WDA950T	W M Buses	WTU485W	Britannia		

ISBN 1 897990 17 0
Published by *British Bus Publishing*
The Vyne, 16 St Margarets Drive, Wellington,
Telford, Shropshire, TF1 3PH
Fax and order-line: 01952 255669

Printed by Graphics & Print
Unit A13, Stafford Park 15
Telford, Shropshire, TF3 3BB

The North & West Midlands Bus Handbook

British Bus Publishing

HANDBOOKS

Also available!

The Leyland Lynx - £8.95
The 1996 FirstBus Handbook - £9.95
The 1996 Stagecoach Bus Handbook - £9.95
The North East Bus Handbook - £9.95
The Yorkshire Bus Handbook - £9.95
The Lancashire, Cumbria & Manchester Bus Handbook - £9.95
The Merseyside & Cheshire Bus Handbook - £9.95
The Scottish Bus Handbook - £9.95
The Welsh Bus Handbook - £9.95
The East Midlands Bus Handbook - £8.95
The Model Bus Handbook - £9.95
The Fire Brigade Handbook - £8.95

Coming Soon

The North and West Wales Bus Handbook - £9.95
The South Wales Bus Handbook - £9.95

Get the best!
Buy today from your transport bookshop,
or order direct from:

British Bus Publishing
The Vyne, 16 St Margaret's Drive, Wellington
Telford, Shropshire TF1 3PH
Fax and Credit Card orders: 01952 255669